FARM STRUCTURES

FARM STRUCTURES

BY

K. J. T. EKBLAW, M.S.

ASSOCIATE IN AGRICULTURAL ENGINEERING, UNIVERSITY
OF ILLINOIS; ASSOCIATE MEMBER OF AMERICAN
SOCIETY OF AGRICULTURAL ENGINEERS

Fredonia Books
Amsterdam, The Netherlands

Farm Structures

by
K. J. T. Ekblaw

ISBN: 1-4101-0438-9

Copyright © 2004 by Fredonia Books

Reprinted from the 1914 edition

Fredonia Books
Amsterdam, The Netherlands
http://www.fredoniabooks.com

All rights reserved, including the right to reproduce this book, or portions thereof, in any form.

In order to make original editions of historical works available to scholars at an economical price, this facsimile of the original edition of 1914 is reproduced from the best available copy and has been digitally enhanced to improve legibility, but the text remains unaltered to retain historical authenticity.

PREFACE

In the preparation of this book it has been purposed to provide a treatise concerning farm structures which will appeal not only to the teacher who desires to present the subject to his students in a straightforward and practical way, but to the progressive farmer who recognizes the advantages of good farm buildings. The popular literature on this subject consists mainly of compilations of plans accompanied by criticisms of more or less value, or of discussions of farmsteads too expensive or impractical to be applied to present ordinary conditions. The elimination of these faults has been among the objects of the author in the writing of this text.

The development of the subject is manifestly the most logical, beginning with a description of building materials, followed by a discussion of the basic methods employed in simple building construction, then presenting typical plans of various farm buildings in which the principles of construction and arrangement have been applied. Descriptions of the more essential requirements in the way of equipment and farm-life conveniences are appended. The illustrations have been prepared with the object of making them truly illustrative and of aid in the understanding of the subject matter which they accompany. Comparatively few building plans are included, since most building problems possess so many local requirements that a general solution is impossible; however, the plans presented are typical, and are so suggestive in presenting fundamental principles that a study of them will aid in the solution of any particular individual problem.

PREFACE

It is not intended that the study of this text will produce an architect; but it is hoped that it will provide the student with a sufficient knowledge of building operations to enable him, with some knowledge of carpentry, to erect his own minor structures and to differentiate between good and bad construction in larger ones.

Acknowledgment is made of the receipt of valuable suggestions and advice from Dr. N. Clifford Ricker, Professor of Architecture in the University of Illinois. In certain parts of the work use has been made of the bulletins published by the United States Department of Agriculture and the various State Experiment Stations, particularly those of Illinois, Iowa, Maine, Maryland, New York, and Wisconsin. Suggestions were also received from numerous commercial firms, especial acknowledgment being due to the following:

Universal Portland Cement Company, Chicago.

James Manufacturing Company, Fort Atkinson, Wisconsin.

Fairbanks, Morse and Company, Chicago.

American Radiator Company, New York.

United States Radiator Corporation, Detroit.

Beatty Brothers, Brandon, Manitoba.

Cosgrove-Cosgrove Company, Philadelphia.

Leader Iron Works, Decatur.

Though care has been taken to prevent the occurrence of mistakes, they will undoubtedly appear, and corrections and suggestions for improvement will be gladly received.

K. J. T. EKBLAW.

UNIVERSITY OF ILLINOIS,
URBANA.

TABLE OF CONTENTS

CHAPTER	PAGE
I. BUILDING MATERIALS	1
1. Wood	1
2. Steel	15
3. Stone	16
4. Brick	23
5. Roofing	29
6. Concrete	31
7. Paint	55
8. Glass	61
9. Nails	63
II. LOCATION OF FARM BUILDINGS	67
III. BUILDING CONSTRUCTION	79
1. Foundations	79
2. Framing	84
3. Walls	92
4. Windows	94
5. Doors	95
6. Floors	96
7. Roofs	99
8. Stairs	102
9. Interior Finish	106
IV. ESTIMATING	112
V. DESIGN AND CONSTRUCTION OF FARM BUILDINGS	118
1. Granaries	118
2. Machine Sheds	123
3. Ice Houses	132
4. Silos	136
5. Poultry Houses	186
6. Swine Houses	202
7. Sheep Barns	212
8. Large Storage Barns	218
9. Dairy Barns	235
10. Horse Barns	252
11. General Purpose Barn	256
12. Farm Residence	257

TABLE OF CONTENTS

CHAPTER		PAGE
VI.	VENTILATION	273
VII.	LIGHTING FARM BUILDINGS	285
	1. Candles	285
	2. Kerosene Lamps	286
	3. Air-gas Lamps	287
	4. Acetylene	288
	5. Electricity	291
VIII.	HEATING FARM HOUSES	296
	1. The Open Fire	297
	2. Fireplaces	298
	3. Stoves	300
	4. Hot Air	300
	5. Steam	302
	6. Hot Water	307
	7. Vacuum	311
	8. Design of Various Systems	311
IX.	FARM WATER SUPPLY	314
	1. Pressure Systems	321
	2. Hot Water Supply	326
X.	PLUMBING AND SEWAGE DISPOSAL	330

FARM STRUCTURES

FARM STRUCTURES

CHAPTER I

BUILDING MATERIALS

Wood

THOUGH the use of wood is becoming a more and more expensive proposition as the scarcity and consequently the cost of lumber increases, it will still continue to be one of the great forms of building material. Steel and concrete in many instances are displacing wood as building materials; the substitution is just, logical, and economical, for the new discoveries and developments relating to steel and concrete which have come with recent years prove their superior value for many purposes. However, when we consider that there are yet standing billions of feet of lumber which can be conserved under proper methods of foresting, we can easily see that wood will always have a definite commercial value.

Structure

Woods suitable for structural purposes are usually called *timber*. Almost the entire amount of timber used in the industries is obtained from that kind of trees known as *exogenous*, or those in which the growth occurs by the formation each year of layers of new wood on the exterior beneath the bark. *Endogenous* trees, or those trees whose increase in diameter is accomplished by the addition of woody matter in the interior, supply only a very small percentage of merchantable woods; some examples of the latter class are the bamboos and palms.

When an exogenous stem is cut across, Figure 3, three distinct parts are visible:

First, the *bark*, a scaly material having a thickness of from $\frac{1}{4}$ to $1\frac{1}{2}$ inches or more, which envelops and protects the wood inside. It has no great commercial value except in certain species; for instance, as in making binding strips, or in dyeing or tanning; some barks have a medicinal value.

Second, the *sapwood*, adjacent to the bark, and from $\frac{1}{2}$ to 4 inches thick; it is generally characterized by a lighter color than the surrounding parts, is softer, and less compact than the inner wood.

Third, the *heartwood*, the central portion, generally distinctly separated from the sapwood. The heartwood is that part of an exogenous stem which possesses strength and durability, and which only should be used where these qualities are requisite.

The development of these parts is accomplished by the absorption, in the spring, of juices from the soil with which the roots come in contact. At first these juices, which are converted into sap, serve to form leaves and new stems. From the upper surface of the leaves moisture is given off by the sap, and carbon is absorbed from the air. After the leaves are full grown, vegetation is suspended until autumn, when the sap in its altered state descends chiefly between the wood and bark where it deposits a new layer of wood. This constitutes the annual ring, and covers all parts of the stem and branches.

As the tree grows older, the inner layers become congested with the secretions peculiar to the tree, and cease acting as sap carriers. Their primary function is now a mechanical one, that of keeping the tree from falling of its own weight or from the force of the wind.

The layers of wood that are formed each year appear as rings on the cross section of the stem, and the age of the tree or a portion of it can be ascertained by counting the number of rings. The width of these rings varies greatly with different trees, being influenced by climatic and by soil conditions. In good white pine the thickness of the ring will be perhaps $\frac{1}{12}$ of an inch, while in the slow-growing long-leaf pine it will be only one half as much. Theoretically, these rings should be uniformly circular in shape, but the shape may be so influenced by internal conditions and by external injuries as to result in great irregularity. This regularity or irregularity in shape of the annual rings has considerable to do with the technical qualities of the timber.

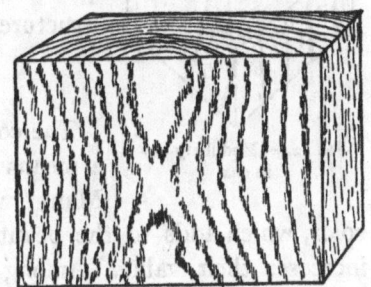

FIG. 1.— Annual rings in pine.

Close examination of the annual rings will show two distinct parts, as in Figure 1, one of which is soft and light colored, designated "spring wood," and another firmer and darker in color, known as "summer wood," from the part of the season in which each was formed. The latter is much the firmer and heavier; consequently, the greater the proportion of it, the greater will be the weight and strength of the timber.

As a whole, wood is made up of bundles of long tubes, cells, or fibers, with their long axis generally parallel to the stem. Cross fibers, known as pith fibers or medullary rays, extend in all directions radially from the pith to the bark, between the linear fibers; the medullary rays act as a binder for the longitudinal fibers. Aside from the linear and cross fibers of woody material, there are resin ducts in pines and spruces, and hollow ducts in the broad-leafed trees. The finished appearance of the wood, and its physical and mechanical qualities, depend greatly upon its structure.

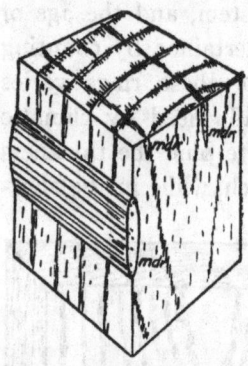

FIG. 2. — Medullary ray in oak.

Color

The color of wood often serves as a means of identifying the species. Many woods have a distinctive color, which adds to the beauty of their appearance, and increases their value. Among these may be mentioned the black walnut, whose heartwood is a beautiful dark brown; ebony, black; cherry, reddish; gum, reddish brown; osage, orange yellow; mahogany, red brown; poplar, light yellow.

Exposure to air or light darkens wood, as is well shown in the case of osage orange, which when freshly cut is a bright yellow; after a little exposure, it becomes a light brown. Color is sometimes an excellent indication of the condition of the wood; the color should be uniform throughout the heartwood in sound timber, the presence of blotches or streaks indicating disease. Decay, dry rot, or fungi cause wood to lose its characteristic translucency, and it is known then as "dead," in distinction to "live" or "bright."

Occasionally woods may be differentiated by their odor. Oaks, pines, cypress, cedar, and apple all have distinctive odors, more or less agreeable. Decomposition is often accompanied by pronounced odors; poplar in decay emits a disagreeable odor, while the red oak becomes fragrant, its smell resembling that of heliotrope.

Defects

Wood is subject to a number of diseases and affections which sometimes materially decrease its value, or even destroy it entirely. Some trees seem to have an age limit, beyond which they gradually deteriorate; others, as the sequoia, or giant redwood of California, continue to grow for centuries without the least apparent deterioration. Felling, handling, and seasoning are important in determining the life of timber, and the methods employed are being so developed as to constitute a science in themselves. Timber felled in winter is more durable than that felled in summer; hewed wood is more resistant to decay than sawed, since the pores are closed and compacted by the blows of the ax, while the saw tears them open.

Dry rot is one of the worst sources of decay to which wood is liable. It is the result of fermentation caused by the spawn of a fungus, upon the introduction of moisture. The wood fibers decay, and the wood crumbles beneath the touch. Wherever there is moderate warmth, dampness, and lack of air, dry rot will occur, and the only means of preventing it is to dry the wood, and apply some substance, either upon the exterior or into the exterior layers, which will prevent the entrance of the fungus.

Wet rot is caused by the presence of moisture, resulting in decomposition of the wood tissues.

Worms cause incalculable damage to wood, especially

FARM STRUCTURES

to that submerged in water. They are known as teredos, or "ship worms," and termites. The teredos are really bivalve mollusks, which bore into submerged timbers to such an extent that heavy timbers are destroyed in four or five years. The termites operate on land, attacking wood after felling, destroying foundation timbers, furniture, etc. Other insects, such as the elm-bark beetle, the oak-borer, etc., also do great damage.

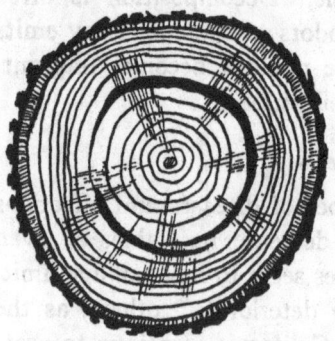
FIG. 3. — Wind or cup shake.

Commercial defects, or defects which aid in the grading of lumber, comprise the following:

Wind-shakes. — Circular cracks separating the annual rings from each other. Figure 3.

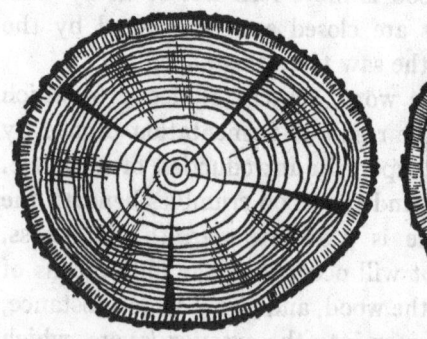
FIG. 4. — Star-shake.

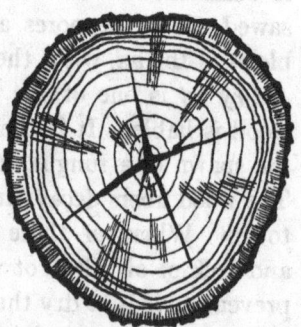
FIG. 5. — Heart-shake.

Star-shakes. — Cracks along the medullary rays, and widening outwards. Figure 4.

Heart-shakes. — Clefts in the center of the log. Figure 5.

BUILDING MATERIALS

Brash. — Timber from trees deteriorating from old age.

Belted. — Timber killed before it is felled.

Knotty. — Timber containing many knots.

Twisted. — Timber in which the grain winds spirally.

Rind-gall. — Swelling caused by formation of layers over a wound.

Upset. — Fibers injured by crushing.

Seasoning

The purpose of seasoning timber is to expel as much of the moisture as possible, thus increasing the resistance to decay, and making it more susceptible to conversion. It is rendered imperative by the changes in volume and shape that all woods undergo under the influence of changes in atmosphere, temperature, and moisture. This is especially important in cabinet and wheelwright work, where stock is usually blocked out and given an extra term of seasoning before using.

The strength of wood is almost always greatly increased by seasoning, consequently it is not economical to use green wood; for it is then not only weaker, but liable to continual changes in volume and shape.

The time of seasoning varies greatly, not only with different kinds of lumber, but with different trees of the same kind, and in different parts in the cross section; there may be twice as much moisture in the sapwood as in the heartwood. Framing lumber is usually stacked in the yards for seasoning, in layers perhaps six feet wide, with inch strips between each layer, so as to permit of the free passage of air. This lumber is rarely seasoned for more than six months before put into use; as a result of this, the shrinkage due to final seasoning causes most of the cracks in the interior of buildings having wooden floors and partitions.

FARM STRUCTURES

Kiln drying is a process of artificial seasoning, in which the wood is put into a tight chamber called a dry kiln, and subjected to a heat of from 150 to 180 degrees F. The time of kiln drying varies from four to five days for soft woods fresh from the saw, to ten days for hard woods which have been air-dried from three to six months previously. Too rapid drying results in "case-hardening," or a formation of a shell on the exterior before the interior has a chance to harden, the later drying causing checks along the medullary rays.

FIG. 6. — Effects of shrinkage.

Effect of Shrinkage. — All woods shrink more tangentially, or along the annual rings, than in a radial direction, because the pith rays resist the radial shrinkage. The effect of seasoning upon a log in different stages of conversion is shown in Figure 6. A log, squared or halved, usually cracks radially, especially near the ends; if cut into boards, the latter take a bent form.

Conversion of Timber

Modern machinery and methods enable the conversion of timber to be accomplished very expeditiously. Gang or circular saws are used to cut the log up into planks or boards, the edges being trimmed off afterward. Edgings are converted into lath or molding, or used for kindling. The United States Forest Service is doing excellent work in saving waste lumber, by investigating the purposes for which scraps ordinarily wasted can be utilized, and bringing the producer and consumer into communication with each other, to their mutual benefit. Even sawdust, huge quantities of which accumulate at lumber mills, is used in the manufacture of wood pulp.

When a whole log is cut up into slices by cuts parallel to each other, the end section of the boards will have the appearance shown in Figure 7; this is known as *bastard sawing*. Except for a few boards cut from the center of the log, the

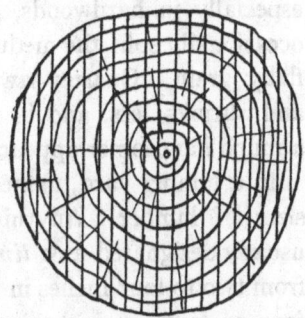

FIG. 7. — Plain or bastard sawing.

end section will show the annual rings in a more or less complete circle. Those cut from near the center will have the annual rings cutting across more perpendicularly to the width of the board, while the medullary rays will extend more nearly parallel to the flat face. Such boards are called *quarter-sawed* from the fact that they are more commonly obtained by first quartering the log and then sawing into boards by cuts at 45 degrees to the flat surface, as shown in Figure 8. When the exposed ends of the annual rings on the flat surface of the sawed board appear perfectly straight, it is known as *comb-grained;* this is especially attractive and valuable in flooring, which receives much wear.

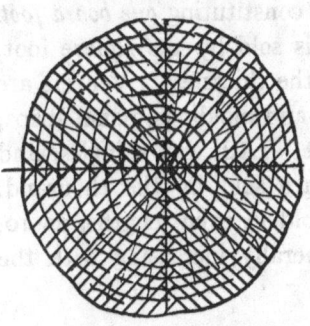

FIG. 8. — Quarter sawing.

Real quarter-sawed lumber, obtained in the manner

described above, is more expensive than the bastard-sawed, on account of the greater difficulty involved in conversion. However, there is a strong and recognized demand for it, especially in hardwoods, as in oaks, where the saw cuts occasionally split the medullary rays, exposing a handsome, flaky grain. Quarter-sawed lumber wears better, warps and shrinks less, and in most of the hardwoods presents a much handsomer appearance than the bastard-sawed.

In a lumber yard, names are arbitrarily given to various sizes of lumber. Anything thicker than four inches is usually designated as a *timber*, or *framing timber*. Lumber from two to four inches in thickness is called *plank*, whether four or fourteen inches wide. *Boards* include all lumber less than two inches thick, of any width, except very thin stuff, which is used for *veneer*. A *veneer* is a very thin strip cut from a log by a veneer machine, there being thirty strips of veneer in an inch thickness of stock.

All large lumber is sold by *board measure*, a board one foot square and one inch thick constituting *one board foot*. Stock less than an inch thick is sold by the square foot, the price varying according to the thickness. Veneers are always sold by the square foot, and moldings, panel strips, etc., are sold by the lineal foot. Lath and shingles, and sometimes weather boarding, are sold by the thousand. In most instances, lumber is sold in even lengths, as 10, 12, 14 feet, etc., the price generally increasing with the length.

Selection of Wood

The user must of course ascertain what kind of wood is most suitable for the work in hand, and then see that the wood he obtains is of the best grade that is consistent with the construction. Even in the same species the lumber will vary in grade considerably, and careful grading is necessary. Winter-felled timber is superior to summer-felled, and the more mature the tree the better will be the timber. Where lightness is desirable, coniferous woods are advantageously used; where jarring loads are to be sustained, denser, tougher woods must be employed.

The following list will aid in the selection of wood for various purposes:

Light framing — white pine, spruce, hemlock.
Heavy framing — yellow pine, oak.
Exterior finish — white pine, poplar, cypress.
Interior finish — redwood, cypress, any hardwood.
Floors — quarter-sawed oak, maple, or hard pine.
Doors and sash — white pine, cypress.
Posts — white cedar, osage orange, cypress, black locust.
Linen closets — southern red cedar.

Testing

The almost universal substitution of iron and steel for the framing of large engineering structures renders exact information relating to the strength of timber less important than formerly. In smaller structures, long experience has so developed the use of proper sizes of timber that one need not worry about safety. A single quality or combination of several may govern the kind of wood used; for example, in a joist, lightness and stiffness are prime requisites; consequently, compressive and shearing strength is subsidiary to transverse or breaking strength; on the contrary, transverse strength may be ignored entirely when wood is being chosen for paving blocks. Where calculations are necessary to ascertain the most economical size or kind of wood to be used, a factor of safety varying from 4 to 6 is employed.

FARM STRUCTURES

REFERENCE TABLES FOR WOOD

CRUSHING STRENGTH

Material	Crushing Weight Lbs./Sq. In.
Cypress	3375
Hemlock	3000
Oak, white	4000
Pine, Georgia yellow	5000
Pine, Oregon	4500
Pine, Norway	3800
Pine, white	3500
Redwood	3000
Spruce	4000
Poplar	3000

SHEARING STRENGTH

Material	Resistance, Lbs./Sq. In. With Grain	Across Grain
Cedar	60	400
Chestnut	125	400
Hemlock	80	600
Oak, white	150	1000
Pine, Georgia	125	1200
Pine, Oregon	125	900
Pine, Norway	90	750
Spruce	90	750
Redwood	70	500
Poplar	60	450

TENSILE STRENGTH
(Factor of Safety, 6)

Material	Tensile Strength Lbs./Sq. In.
Ash, white	2000
Chestnut	1500
Hemlock	1500
Oak, white	2000
Pine, Georgia	2000
Pine, Oregon	1800
Pine, Norway	1600
Pine, white	1400
Redwood	800
Spruce	1600
Poplar	1200

BUILDING MATERIALS

WEIGHT

Material	Weight in Lbs /Cu Ft.
Cypress	34
Elm	35
Hemlock	25
Hickory	52
Mahogany	53
Oak, white	48
Pine, white	25
Pine, Georgia yellow	45
Poplar	29
Spruce	25
Walnut	38

Common Varieties of Timber

Ash, white — color, brown; sapwood, lighter, often white; wood, heavy, hard, strong, coarse grained; use, interior work, cabinet work, implements.

Cedar, white — color, light brown; thin sapwood, nearly white; wood, light, soft, rather coarse grained; use, posts, ties, shingles.

Cedar, red — color, reddish brown; wood, light, soft, brittle; use, interior finish, shingles, storage chests.

Cypress — color, bright, light yellow; wood, light, hard, brittle; close grained; use, interior finish, posts, sills, cabinet work.

Elm, white — color, clear brown; wood, heavy, hard, tough; use, posts, bridge timbers, sills, ties.

Gum — color, bright reddish brown; wood, heavy, hard, tough, close grained, inclined to shrink and warp; use, interior and exterior finish.

Hickory — color, brown; more valuable sapwood, white; wood, heavy, very hard and strong, close grained; use, implements, vehicles.

Hemlock — color, bright brown to white; sapwood, darker wood, light, soft, weak; use, rough framing.

Locust — color, brown; sapwood, yellow; wood, heavy hard, strong; use, posts, turnery.

Maple, hard — color, light reddish brown; wood, heavy hard, strong, tough, close grained; use, flooring, interior finish.

Maple, white — color, white; wood, hard, strong, brittle; use, flooring, interior finish.

Oak, white — color, brown; wood, heavy, hard, strong, close grained; use, interior finish, furniture.

Oak, red and black — color, light brown or red; wood, heavy, hard, coarse grained; use, interior finish and furniture.

Pine, white — color, light brown, often tinged with red; wood, light, soft, very close, straight grained; use, interior finish, windows, doors, etc.

Pine, Norway — color, light red; wood, light, hard, coarse grained; use, in all construction work.

Pine, yellow (long-leaf) — color, light red or orange; wood, heavy, hard, strong, coarse grained; use, in all construction work; the inferior short-leaved yellow pine is often substituted for it.

Pine, Oregon (Douglas Fir) — color, light red; wood, hard, strong; use, in construction work.

Poplar (white-wood) — color, light yellow; wood, soft, brittle, very close, straight-grained; use, interior and exterior finish.

Redwood — color, clear, light red; wood, light, soft, brittle, coarse grained; use, building material.

Walnut, black — color, rich dark brown; sapwood, much lighter; wood, heavy, hard, strong; use, interior finish, furniture.

STEEL

The chemical composition of steel is intermediate between *cast iron* and *wrought iron*. It is a mixture of ordinary iron with small varying quantities of carbon, and is produced either by partly eliminating carbon from pig iron, or by adding carbon to wrought iron. Steel, by virtue of its carbon content, possesses the property of tempering; however, in very soft steels, in which the carbon is low, tempering is not apparent.

Classification

Commercially, steel is designated as "*mild*" or "*soft*," "*medium*," and "*hard*." Though there is no definite line of demarcation, generally it is taken that soft steels contain less than 0.15 per cent carbon, that hard steels contain above 0.30 per cent, while the intermediate grades constitute medium steel. The tensile strength of steel varies with the percentage of carbon, from an ultimate strength of 54,000 pounds per square inch for a carbon content of 0.08 per cent, to 87,000 pounds per square inch for a carbon percentage of 0.40.

Method of Manufacture

Almost the entire output of steel used for structural purposes is made either by the *Bessemer* process or by the *open-hearth* process. Small quantities of higher grade steel are made by special methods.

Bessemer Process. — This process involves the melting of pig iron, containing a large percentage of carbon, in a cupola, or running it direct from a blast furnace into a converter, a pear-shaped vessel lined with fire brick. While the metal is in the converter a strong blast of air is forced through it until all the carbon is removed. Then a small quantity of spiegeleisen, a metal containing a known percentage of carbon, is added, in order to control the carbon content of the steel, and when the two metals are thoroughly incorporated, the steel is emptied into molds.

Open-hearth Process. — In this process pig iron and ore are melted together on the open hearth of a regenerative gas furnace. The pig iron is first melted and raised to a temperature which will melt steel; then rich, pure ore and limestone are added. The chemical reactions result in the formation of a fusible slag composed of the lime and silicic acid formed from the silicon, and the carbon passes off as carbonic acid, leaving the steel clear. In general, the open-hearth process results in a better quality of steel, though more expensive, since it takes more than ten times as long to manufacture as by the Bessemer process.

Properties

Steel is slightly heavier than wrought iron, weighing about 490 pounds per cubic foot. Its melting point varies from 2372 degrees to 2687 degrees F., according to the percentage of carbon. The strength of steel varies considerably with its chemical constituents, carbon, manganese, silicon, phosphorus, and sulphur being the main elements affecting it.

Tensile Strength 0.25% C. . . . 68,200 lbs. / sq. in.
Shearing Strength . . . . 10,000 lbs. / sq. in.
Crushing Strength . . . . 48,000 lbs. / sq. in.

STONE

Classification

Three classifications are necessary to give a fairly complete comparison of the rocks from which stones used as structural material are selected; namely, *geological*, *physical*, and *chemical*.

Geological Classification. — Nothing very definite can be determined as to its relation to the properties of stone as a building material from this classification, except that generally the older the rock formation, the stronger and more resistant to the action of the elements and to the effect of deterioration is the stone; but this is not true in many cases. Various things make exceptions to this; for example, a seismic disturbance may so loosen the structure of the rock as to totally destroy its crushing strength. Classifying rocks according to their origin, we have three classes:

Metamorphic, constituting the class including granite, slate, marbles, etc.

Igneous, of which basalt, lava, and trap are examples; and

Sedimentary, comprising sandstones, limestones, and clay. Clay may not be strictly considered a stone, but since it enters into the manufacture of artificial stone, it may be properly included in the classification.

Physical Classification. — In considering the structure of rocks as they exist in natural masses, two great divisions may be observed, *stratified* and *unstratified*.

Stratified rocks, in which the structural material is composed of sheets or layers of varying thicknesses, superimposed upon each other, originally horizontally, which arrangement may at later times have been changed by disturbing forces to a vertical, inclined, or curved position. These layers have evidently originated from depositions from water, from the fact that the rock can be more easily split at the planes of division between adjacent layers than at any other place. The structure of stratified rocks governs to a great degree the method of placing, for besides at the principal planes they may be more or less easily split at intermediate planes, which may lie parallel, oblique, or perpendicular to the principal planes, and which divide the rock into laminæ. Laminated stones should be placed in buildings so their *laminæ* or *beds* are perpendicular to the direction of greatest pressure, and with the edges of the laminæ as nearly as possible perpendicular to the exposed surface of the stone; their crushing strength is greatest parallel to their laminæ, and their durability is greatest with only the edges of the laminæ exposed to the weather.

Unstratified rocks include those composed of crystalline materials held together by some more or less firm cementitious material, which under the influence of decay or heavy shock loosens and leaves the crystals separate. There seems to exist in large masses of unstratified rock an occasional weak streak, these streaks having apparently the same direction. This fact enables quarrying and cutting to be accomplished more readily if advantage is taken of the "rifts" or "lines of cleavage."

Chemical Classification. — By determining the predominant mineral in the chemical composition of rocks, three classes are differentiated:

Siliceous, stones whose base is silica, of which granite and trap rock are examples.

Argillaceous, comprising clay, shale, slate.

Calcareous, represented by limestones and marbles.

Varieties of Stone Masonry

Ashlar Masonry. — This type of stone masonry consists of blocks of stone cut to regular figures, generally rectangular, and built in courses of a uniform height or rise, which is seldom less than a foot. The softer stones should have a length greater than three times the height or rise, while in harder ones the length may be four or five times the depth. In laying ashlar masonry, no side joint in any course should be directly above a side joint in the course below, but the stones should overlap or break joints to an extent of from one to one and one half times the depth of the course. In this way two stones support the weight of the one below, the pressure is more equally distributed, and the structure is bonded together.

Broken Ashlar. — This consists of cut stones of unequal depths, laid in the wall without any attempt at maintaining courses of equal rise, or stones in the same course of equal depth. As in ashlar masonry, one fourth of the face of the wall should consist of headers.

Squared Stone Masonry. — This differs from the regular ashlar in the character of the dressing and the closeness of the joints. The stones are only roughly squared and roughly dressed on beds and joints, so that the width of the joint is half an inch or more, instead of one fourth of an inch, as in ashlar.

Rubble Masonry. — Masonry composed of unsquared stones is called rubble. It comprises two classes: (1) *uncoursed rubble*, in which irregular stones are laid without any attempt at coursing; (2) *coursed rubble*, in which the blocks of unsquared stone are leveled at specific heights to an approximately horizontal surface. These courses may not be of the same depth, in which case they are known as *random courses*. In building rubble masonry, weak angles should be knocked off the block, each stone should be cleansed from dust, dirt, etc., and each one must be firmly embedded in mortar, resting as far as possible on its largest side.

Measurement of Stone

There is not an established rule for the measurement of masonry, it being governed largely by local custom. Stone may be bought under two classifications, *rubble* and *dimension stone*. The former comprises the rough, irregular blocks of stone which cannot be laid in courses or with square joints without further trimming. The latter includes the stones cut to size, usually with a rectangular face, or in a shape in accordance with the architect's specifications. The price of the dimension stone is, of course, considerably higher than rubble, on account of the work in cutting.

In general practice, rubble is sold by the carlot or by the ton, while dimension stone is sold by the cubic foot. When it is cut to some particular thickness and is to be used for some specific purpose, as for footings, facing, or flagging, it may be sold by the square foot.

In stone masonry, rubble work is almost universally measured by the *perch* of $16\frac{1}{2}$ cubic feet, though a legal perch is 50 per cent larger. In large structures, such as massive foundations and bridge piers, the cubic yard is the unit of measurement. Sometimes, in architectural details, the unit of measurement is the superficial square foot.

Methods of cutting Stone

Cut stone includes all squared stones, with smooth beds and joints. There are a great many ways in which the face may be finished, but the following are the chief ones (shown in Figure 9):

a. Rough pointed, with projections varying from $\frac{1}{2}$ inch to 1 inch, made by heavy picking.

BUILDING MATERIALS

b. Fine pointed, a smoother finish obtained by following rough pointing with a lighter, finer tool.

c. Crandalled, a variation of fine pointing, using a tool with a toothed edge.

d. Axed, the face appearing covered with parallel chisel marks.

e. Bush hammered, a fine-pointed finish, smoothed with a bush hammer.

f. Rubbed, a finish obtained by rubbing with grit or sandstone.

g. Diamond panel, a very flat pyramid cut inside the margins, with its apex at the center of the block.

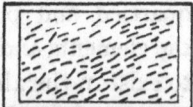
Rough Pointed

Fine Pointed

Crandalled

Axed.

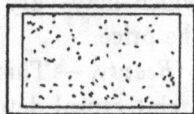
Bush Hammered.

Diamond Panel

FIG. 9 — Methods of cutting stone.

Requisites for Good Building Stone

Four essentials are present in all good building stone; namely, *durability*, *strength*, *cheapness*, and *beauty*. These may vary in degree of importance for different structures, and for different parts of the same structure. For example, limestone may be employed both in the foundation and in the superstructure of a building; for the former, those stones whose appearance is in any way marred by flaws in texture, shape, or finish should be used, leaving the better ones for the superstructure, where a handsome appearance is a prime consideration.

Durability. — This quality of building stone is one about which only suppositions can be made. Though the rock may have existed in a natural state for centuries, yet when quarried and put in a building, exposed to the action of the elements and to the action of the acids which are known to exist in the air, deterioration or even decomposition may occur with remarkable rapidity.

Strength. — This is an indispensable quality, and one which governs the selection of stone for foundations, piers, etc. The resistance to crushing, in pounds per square inch, varies from 5000 for some Maryland granites, to 43,000 for Potsdam (N.Y.) red sandstone. As a rule, however, granites are stronger than sandstones and limestones. The crushing strength of any stone should be thoroughly investigated before it is used in structural work.

Cheapness. — The cost of stone is governed by various factors, chief among which are suitability for structural purposes, accessibility to market, and the ease with which it may be quarried and dressed. The quality of the stone, which may vary greatly in the same quarry, also affects the price.

Beauty. — This quality is mainly considered where architectural details are involved, and where the stone is employed in sculpture.

Varieties of Building Stone

Trap is the strongest of all building stones, though because of its toughness and difficulty of conversion it is little used. For macadam and railroad ballast it is unexcelled.

Marble, in common practice, is any limestone that will take a good polish. In reality, marbles are only those limestones which have undergone metamorphic action in which their texture has become more crystalline, and their color modified. Marble is the most beautiful of building stone.

BUILDING MATERIALS

Granite is the strongest and most durable of building stones in common use. It is readily quarried and converted, and can be used for almost any purpose. Granites vary in color from gray to black, and from pink to red.

Limestones. — The chief constituent of limestones is carbonate of lime, though the varieties are exceedingly numerous. They are quite durable and strong, and because of the pleasing colors they are esteemed for building purposes.

Sandstones are more easily worked under the chisel than limestones, and for this reason and because of their abundance are more generally used. They are composed chiefly of sand, more or less cemented and consolidated, and their texture varies with the coarseness of the sand. Sandstone becomes harder and more durable with age, on account of the precipitation of soluble silica. In color, sandstone shows the widest variation, almost every shade existing from white to black, including dark red, and even purple.

BRICK

Brick is an artificial stone made by subjecting an intimate mixture of clay or shale and water, molded into shape, to a heat intense enough to cause partial vitrification. Different kinds of clays and different methods of preparation result in different qualities of brick. Various chemical constituents in the clay affect the finished product to a considerable extent. Iron gives hardness and strength to brick, and gives it a characteristic red color. Silicate of lime makes clay soft and too fusible. Alkalies are found in small quantities in the best of clays, and are more or less detrimental. Sand is sometimes included in the mixture to prevent excessive shrinking in burning. The manufacture of brick includes the excavation of the clay or shale, the preparation of the clay by tempering and molding, the drying of the molded mixture to drive off superfluous moisture, and the burning of the molded mixture in kilns.

Classification

There are three bases for classification of bricks: 1. The method of molding. 2. Position in kiln when being burned. 3. Their shape or use.

The first classification gives the following:

Soft-mud brick, molded from a soft mixture of clay and water.

Stiff-mud brick, molded from a stiff mixture.

Pressed brick, made from dry or semi-dry clay under heavy pressure.

Repressed brick, a *soft-mud* brick, subjected to enormous pressure when partially dry to give it a better form and texture.

The second classification includes:

Arch or *clinker* bricks, those which form the tops and sides of the arches of brick around the fires. They are vitrified, hard, and usually weak from overburning.

Body or *hard* bricks, those taken from the interior of the pile, and best in quality.

Salmon or *soft* bricks, forming the exterior of the pile, and too soft from underburning to be of great value.

The shape or use of brick gives rise to the following classification:

Compass brick, those having one edge shorter than the other, used in shafts, cisterns, etc.

Feather edge or *voussoir*, those having one edge thinner than the other, used in arches.

Front or *face brick*, those made especially for facing the walls of buildings.

Sewer brick, ordinary hard, smooth, regular brick.

Kiln-run brick, brick made from fire clay, sufficiently free from alkali silicates, iron, and lime to resist vitrification at high temperatures; used in furnaces, cupolas, fireplaces, ovens, etc.

Measurement of Brick

The methods of measuring brickwork are even more arbitrary and farther from being standardized than are those of stone masonry. The *perch* is sometimes used, though it varies in different localities from $16\frac{2}{3}$ to 25 cubic feet. In some lines of work the unit is the superficial square foot, allowing a certain number of brick for various thicknesses of wall. The cubic foot or the cubic yard is unsatisfactory as a unit because of the variation in the size of brick. The most satisfactory method is to specify that the masonry will be paid for by the cubic foot or cubic yard, according to the arrangement between architect and contractor.

Size, Weight, and Strength

The *size* of brick is a matter of such indefiniteness that brick of almost every conceivable size can be found. There is no legal standard. Eastern bricks average about $7\frac{3}{4} \times 3\frac{3}{4} \times 2\frac{1}{4}$ inches. Western bricks average about $2\frac{1}{2} \times 4\frac{1}{8} \times 2\frac{1}{2}$ inches. The size of all common bricks varies even in the same lot, according to the degree of burning, the hard bricks being $\frac{1}{8}$ to $\frac{3}{16}$ of an inch smaller than the salmon brick.

Pressed and faced bricks are generally more uniform in size, being made in more uniform molds, and on account of their density being less affected by burning.

The *weight* of bricks is controlled by the quality of the

clay from which they are made, by their size, and by the density resulting from pressing. Ordinary bricks weigh about 4½ pounds, while pressed bricks weigh from 5 to 5½ pounds each.

The *strength* of brick is also a matter of great variation; ordinary brick have a crushing strength of about 6000 pounds per square inch, while pressed brick reach 13,000 pounds per square inch.

Joints

Much of the strength and appearance of brick masonry depends upon the care exercised in laying the brick. The thickness of the mortar joints should not be over ⅜ of an inch, ¼ inch being the commonest thickness. When good brick are used, the mortar is the weak part of the wall; consequently the less used of it, the better. Mortar should be spread as thinly as is consistent with a uniform bearing and rapid work in spreading the mortar. Especially is this true with exterior walls, where the mortar is subjected to disintegration by the elements.

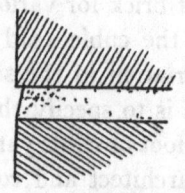

FIG. 10.—Facing a mortar joint. This method prevents weathering.

What in common practice is called a "shove-joint" results in well-laid brick. The brick is first laid so as to project over the one below, then pressed into the mortar and shoved back into its final position. In order to prevent disintegration of mortar in exterior work, extra lime is added to mortar for face brick; the mortar may contain some cement, or the joint may be pointed with neat cement mortar. If the joint is faced as shown in Figure 10, moisture will not penetrate to such an extent as in a square face.

BUILDING MATERIALS

Bond

The arrangement of the bricks in the surface of a given wall according to a system is known as the *bond* of the brickwork. In order to make brick masonry have sufficient strength, it is necessary to have the bricks overlapping both across the wall and along the length of it, and in devising arrangements to meet this condition, several varieties of bond have been originated. Sometimes, instead of laying bricks crosswise in a wall, the adjacent vertical layers are held together by a bond made of a piece of wire or strap iron bent up slightly at the ends, and embedded in the mortar between the bricks.

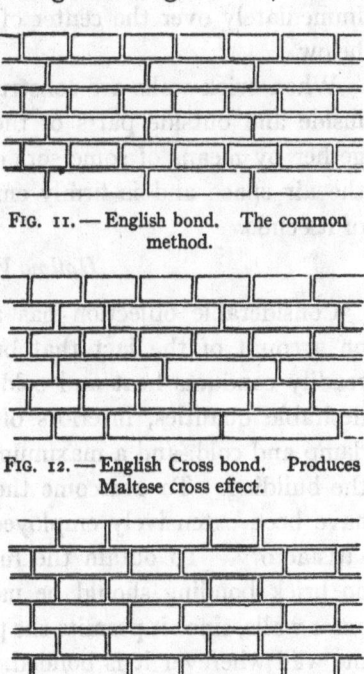

FIG. 11. — English bond. The common method.

FIG. 12. — English Cross bond. Produces Maltese cross effect.

FIG. 13. — Flemish bond. A simple but attractive method.

The most widely used bonds of the self-contained type are the *English* and the *Flemish* bonds. The former consists of entire courses of headers and stretchers; a header being a brick laid with its end exposed in the face of the wall, a stretcher having its edge exposed. Sometimes the courses of headers and stretchers alternate, but more usually the proportion of courses of headers to courses of stretchers is one to from four to six, as shown in Figure 11. A variety of this, known as the *English Cross* bond, has alternate courses of headers and stretchers, the stretchers of the successive stretching courses breaking joints, as in Figure 12. This makes an attractive bond.

Flemish bond, shown in Figure 13, consists of alternate headers and stretchers in every course, every header being immediately over the center of the stretcher in the course below.

When brick walls are constructed with an air space, the inside and outside parts of the wall must be bonded together by means of some sort of a metal *tie* which crosses the air space and is firmly embedded in mortar at both of its ends.

Hollow Walls

Considerable objection has arisen to solid brick walls on account of the fact that brick absorbs moisture, and readily conducts heat and cold. As a result of these undesirable qualities, interiors of houses built of brick are damp and cold, and a maximum of fuel is required to heat the building. To overcome these objections, hollow walls have been extensively employed, and seem to be entirely satisfactory. To obtain the full benefit of the air space, no brick bonding should be used between the inner and outer walls, since it permits the passage of moisture through the wall wherever it is bonded. At the bottom of the air space some means should be provided to drain off the moisture which permeates the outer wall and drops to the bottom.

Efflorescence

The face of brickwork is sometimes discolored to a greater or less degree by a white *efflorescence* which appears after the bricks have been laid, and which may reappear years afterward following a driving rainstorm or damp snow. It is caused by one or more of several things: the action of the lime in the mortar upon the silicate of soda in the bricks, or the union of the magnesia in the lime mortar with the sulphuric acid formed by burning clay containing pyrites. The efflorescence is never due to any constituent of the mortar or the bricks alone, and methods are now being perfected to prevent it. Any preparation which, when applied to the surface of the brick, will make them impermeable to moisture will prevent the efflorescence from appearing.

Roofing

Shingles as roof covering are used far more than any other type for residences, farm buildings, sheds, etc. The best shingles are made from cypress, redwood, or cedar, in the order given. Cypress shingles are usually 18 inches long and are supposed to be $\frac{7}{16}$ of an inch thick at the butt, while other kinds are but 16 inches long and about $\frac{5}{16}$ of an inch thick at the butt. The width of shingles varies from $2\frac{1}{2}$ to 14 and even 16 inches. They are sold in bundles, usually four to a thousand, a "thousand" meaning the equivalent of 1000 shingles 4 inches wide. When shingles are to be used for special designs, they are sawed to a uniform width, either 4, 5, or 6 inches, and are known as dimension shingles.

Slate shingles are used where fireproofing and permanency are of importance. A good slate should be hard, tough, and uniform in quality and color. The color of slates varies from blue-black, dark blue, and purple to gray and green, and in some quarries, red. The size of slates is also subject to variation, from 6×12 inches to 14×24 inches. They are sold by the "square," which means a sufficient number of slates to cover 100 square feet of roof with a 3-inch lap over the course below.

Roofing tile is a term applied to exterior roof covering, made from clay, with overlapping edges. Their comparatively high cost has prevented the wide use of tile in America, though in better classes of residences their use is common because of their adaptability in lending themselves to fancy treatment in architectural details. They compare favorably with slates in cost. Tile manufactured from sheet metal heavily tinned or galvanized, or painted, are coming into quite common use.

Tin roofing is made with the use of sheets of steel coated with tin or a mixture of lead and tin, called *terne*. Where the roof pitch is less than one third, the plates are united with flat seams, and are fastened by means of one-inch tinned and barbed roofing nails over which the seams are well hammered down, and then soldered. For steep roofs, standing seams should be used composed of two "upstands" with a cleat holding them in place, as shown in Figure 14. Nails should be driven into the cleats only. A tin roof properly made and kept well painted should last thirty or forty years.

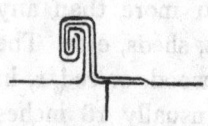

FIG. 14. — Standing joint on a tin roof. When pressed together this makes a very tight joint.

Gravel roofing is used on very low-pitched roofs. It is formed ordinarily by covering the surface of the roof with dry felt paper, and over this laying three, four, or five layers of tarred or asphalted felt, the layers overlapping each other, so that only from 6 to 10 inches of the 30-inch width of paper is exposed. This is then covered with a uniform coat of pitch into which, while hot, gravel or slag is imbedded. A responsible roofer will usually guarantee his work for five years, although a good roof of this kind should last from fifteen to twenty years.

"*Ready roofing*," made by cementing together two or more layers of saturated felt or felt and burlap, and then coated with either a hard solution of the same cementing material, or with hot pitch or asphalt in which is imbedded sand or fine gravel, is quite widely used. It is usually sold in rolls 36 inches wide. When made by a reliable manufacturer, it provides an economical and durable roof, and for some buildings it is to be preferred to any other form of roofing.

CONCRETE

Concrete is a mixture of *water*, hydraulic *cement*, and an *aggregate* composed of sand, gravel, or broken stone, in certain definite proportions, which, when allowed to harden, form an artificial stone.

The use of concrete is as old as history itself. The huge temples of Babylonia and Assyria, with their enormous columns and arches, were built of concrete; so were the Aztec and Toltec temples of Mexico and South America; the Romans used concrete extensively in their large public building; even the Pyramids of Egypt, the construction of which is a source of marvel to engineers, are now claimed to have been built of large concrete blocks, probably cast in place.

Cement, though so widely used by the ancients, seemed to fall into disrepute during the Middle Ages, and lime and silt mortars were used instead. Many of the famous cathedrals of Europe were begun at this time, and the inefficacy of this form of construction is seen in the fact that these structures have been constantly repaired since their building began, the mortar joints disintegrating and requiring refilling and repointing.

About the beginning of the eighteenth century, however, the use of true hydraulic cement began, and Portland cement, so named by its originator from the resemblance it bore to the stone from the Portland quarries in England, became a great factor in all kinds of construction. In 1912, nearly 80,000,000 barrels of it were manufactured in the United States alone.

Concrete Materials

Cement. — For large structures, where the amount of cement used may run into hundreds or thousands of barrels, the selection and testing of the cement are of vital importance. Standard specifications have been evolved by the American Society for Testing Materials, and the cement must fulfill these rigid requirements before it is accepted. For small structures, and especially for the small pieces of concrete work on the farm, cement need not be tested. If it is of a standard brand, and if it is bought of a reliable dealer, its worth should be sufficiently assured to warrant its use. In storage it should be kept dry, for the presence of even a small amount of moisture will cause it to harden, and then it cannot be used again.

Sand. — Ordinarily concrete is mixed in certain proportions of cement, sand, and gravel. Since the proportions depend a great deal upon the sizes of the aggregate, the selection of it is important. The sand should be clean, and consist of particles of varying sizes, in order that voids may be eliminated as completely as possible; the larger the voids, the more cement must be used. This is illustrated in Figure 15; where the aggregate is composed of particles of all sizes, the total voids are less than a third of what they are in an aggregate of uniform size. Clay or loam in sand is an undesirable constituent; some au-

BUILDING MATERIALS

thorities claim that a small percentage is not injurious, while others claim it is; at any rate, it can do no harm to have the sand clean.

Gravel. — The same precautions to be observed in the selection of sand apply equally well for gravel. The maximum size of the gravel particles depends upon the purpose for which the concrete is to be used; fence posts, for instance, requiring the coarse aggregate to consist of particles not larger than the end of one's finger, while for

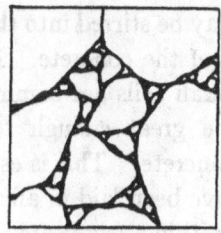 

FIG. 15. — Varying and uniform aggregate. Showing economy of the former.

mass work, as foundations, large bowlders, well embedded, may be used to advantage without weakening the concrete.

Broken stone. — This is used for the coarser aggregate in many instances where gravel is not available. The comparative value of gravel and broken stone as aggregates for concrete has ever been a bone of contention, but for small structures such as are made on the farm, it need not be taken into consideration. Broken stone is usually crushed limestone, which has been graded according to size, from screenings up to pieces which will just pass through a 2½-inch ring. This uniformity in size is undesirable, because it detracts from the void-filling properties of a correctly constituted aggregate. To obtain the best results, where stone must be used, it should be obtained in different sizes, in such proportions that when thoroughly mixed a mimimum of voids will exist. Screenings, or stone dust, is a valuable material for making the finish coat on sidewalks, or in any sort of concrete work where the details must be brought out so carefully as to preclude the use of coarse aggregates.

Water. — The only precaution to observe in using water to temper concrete is to be sure that the water is clean, and not alkaline. Sometimes, in the construction of retaining walls, abutments, etc., water taken out of the stream is used, and enough silt may be stirred into the water to seriously impair the strength of the concrete. In some sections of the country where alkali soils are common, the alkalinity of the water may be great enough to cause ultimate disintegration of the concrete. This is especially the case with drain tile which have been laid in alkali soils, and the resulting disintegration within a few years has done much to discourage the use of concrete drain tile, even in soils where no alkalinity exists.

Mixing Concrete

The materials for making concrete may be mixed either by hand or by machine, the latter method being used universally for large jobs. Hand mixing is done less generally than formerly even in small jobs because users of cement are recognizing the value of the time lost in hand mixing, and because small batch or continuous mixers can be purchased at a low cost; even simple ones can be made at a small expense.

When hand mixing is employed, the cement and sand should first be mixed dry until the two materials are thoroughly incorporated; then the dry mixture should be tempered with water until it is of the proper consistency, the coarser aggregate of screened gravel or broken stone being added as the mixing proceeds. The various degrees of consistency may be arbitrarily classed as *dry*, *medium*, and *wet*. A dry mixture is one whose degree of dampness is about the same as that of damp soil; only heavy, continued tamping will expose water. A medium mixture is so wet that it will barely hold its shape when heaped up. A wet, or sloppy, mixture is one similar in consistency to mortar for plastering.

Proportioning

Concrete materials are mixed in some stated proportions when only a small amount is required; for example, one part of cement, two parts of sand, four parts of gravel; these parts are always measured by volume, and such a proportion is known as a 1:2:4 mixture. Taylor and Thompson, in "A Treatise on Concrete," arbitrarily make four proportions which differ in relative quantities of cement and which serve as a guide to the selection of amounts of materials for various classes of work. These proportions are as follows:

1. Rich — $1:1\frac{1}{2}:3$; for columns and structural parts subjected to heavy stresses.

2. Standard — $1:2:4$; for floors, beams, and columns requiring reënforcing, for tanks, sewers, etc.

3. Medium — $1:2\frac{1}{2}:5$; for walls, piers, sidewalks, etc.

4. Lean — $1:3:6$; for heavy mass work which is only in compression.

These proportions, however, are not theoretically correct. The determination of the exact percentage of voids is a very technical process, and usually is not done, except in large structures, where the aggregate used is fairly constant is size. The ideal proportioning is accomplished when the percentage of voids is reduced to a minimum, the finer aggregate just filling the voids between the particles of the coarser aggregate, and just enough cement being used to coat thoroughly every particle of the whole aggregate and to fill the minute remaining voids.

For determining the amount of materials in a cubic yard of concrete, the following formulas, known as Fuller's Rule, give fairly accurate results:

Let c = number of parts of cement
s = number of parts of sand
g = number of parts of gravel or broken stone

Then

$P = \dfrac{11}{c+s+g}$ = number of *barrels* of cement required for *one cu. yd.* concrete.

$P \times s \times \dfrac{3.8}{27} = S$ = number of *cu. yd.* sand required for one cu. yd. concrete.

$P \times g \times \dfrac{3.8}{27}$ or $S \times \dfrac{g}{s} = G$ = number of *cu. yd.* gravel or broken stone required for one cu. yd. concrete.

When making mortar, it is generally assumed that the following table holds:

Parts Cement	+ Parts Sand	= Parts Mortar
1	1	1.4
1	2	2.2
1	3	2.8

Special Properties of Concrete

As a building material, concrete has been subjected to some extremely difficult tests which have proved its worth. Sea water, however, seems to have a destructive action upon it, the dissolved sulphates forming acids which decompose the cement. By controlling the composition of the cement and keeping the lime and alumina content as low as possible, this decomposition may be more or less preventable; the imperviousness which results from the use of a rich mixture will also preserve the structure.

Effect of Temperature. — Freezing and thawing have practically no effect on concrete. Freshly laid concrete, however, is sometimes seriously injured if it freezes before it sets, and the laying of concrete in freezing weather should be avoided. When it is necessary to do it, both the water and the aggregate should be well heated before mixing, and the concrete should be laid rapidly. If practicable, it should be protected by canvas, clean straw, or some such material. There seems to be a certain amount of heat generated in the setting of cement, which, when retained, will keep the concrete sufficiently warm to enable it to set properly.

Fire and Rust Protection. — Experiments and observations have conclusively proved that concrete is an admirable protection of steel from both fire and rust. A coating of dense concrete, $1\frac{1}{2}$ or 2 inches thick, made in ordinary proportions with gravel or cinders, will resist the most severe fire likely to occur in any building, and will prevent the corrosion of steel even under extraordinary condition.

Water-tightness. — Though concrete when mixed in lean proportions is more or less porous and readily absorbs moisture, it may easily be made water-tight. This is accomplished in several ways:

1. By ideal proportioning of the cement and aggregates.
2. By special surface treatment.
3. By intimately mixing with the concrete some foreign substance which prevents the absorption or passage of water.
4. By applying asphalt and felt, or other waterproof material.

The consistency of the mixture, too, has an effect on the waterproof qualities of concrete, it being found that the wetter the mixture up to a certain degree, the more impervious will be the concrete.

The two methods first listed above are not especially efficient; the fourth method has been much used in the past and is decidedly effective, though very expensive. Modern practice in waterproofing concrete tends to the use of the third method, the market being flooded with waterproofing mixtures of more or less merit. They are comparatively cheap, and for certain classes of concrete work are very effective and valuable.

Forms

A very important consideration in concrete construction is the form. Concrete is a plastic substance, and reproduces with fidelity every detail of the cavity into which it is put. Consequently, the preparation of the forms cannot be slighted. Various materials are used in form construction, heavy foundations below ground having no other form than an earth wall, but the greatest number of forms are made of iron or wood. Iron has its advantages, since it can be easily cleaned and can be used an indefinite number of times for the same work; but its use is limited, on account of the difficulty in working it. Wood forms are used very extensively, because wood is easily convertible, and can almost always be obtained in sufficient quantities.

Green spruce or fir is a suitable wood for forms, for it will not warp, and does not absorb moisture to any great degree. If wood is to be used over and over again, it is economical to use the best-grade matched stuff, free from loose knots, and to have it built up in as large sections as it is practicable to handle. The cement will gradually fill up the pores of the wood, and thus preserve it as well as would a coat of paint.

Before filling the concrete into the forms, paint them with some greasy mixture, oil or soap. This is to prevent the forms from sticking to the cement. Should any particles of cement adhere to the forms when they are removed, they should be immediately and completely removed.

Surface Finish

Unless exceedingly great care is taken in the preparation of forms, the surface of the concrete will present an unattractive appearance. To remedy this, various methods of surface treatment are resorted to. A cement wash, composed of a neat mixture of cement and water, may be spread over the surface, and will leave a clear, smooth exterior unless the surface is exceedingly rough. In this event, the concrete is gone over with a solution of dilute hydrochloric acid, and scrubbed with a wire brush, to remove the surface cement. Then a coat of cement plaster is applied, either smooth, cast, or rubbed, this hiding the comparatively lifeless surface of the bare concrete.

Ornamental surfaces may be obtained on concrete by brushing, rubbing, tooling, or by using an aggregate of some attractive color.

Where the finish is to be obtained by brushing, the forms must be removed as soon as possible and the brushing accomplished rapidly while the cement is still green. Care must be taken that it is not done too soon, as little particles of the aggregate will be loosened, resulting in a pitted and unsightly surface. A brush with stiff, springy bristles, either fiber or wire, will serve the purpose if the cement does not get too hard, and a liberal use of water will materially assist in the work.

A rubbed concrete finish is obtained by removing the forms when the concrete is a day or two old, and rubbing the surface with some abrasive material, such as emery, sandstone, etc. To get the best results from this treatment, the aggregate used in the concrete should be rather fine, or if there is any coarse stuff, it should be spaded back from the face of the work. To assist in getting a smooth surface, a grout of cement should be worked into any little existing crevices. This surface treatment erases form marks, and is superior to painting with cement wash, since there is nothing to scale off.

Tooling may be done on concrete just as effectively as on stone, provided the surface of the concrete has no large aggregate in it. To obtain the best results, the concrete should be thoroughly hardened before any work is attempted on it.

Almost any color and texture can be obtained by choosing for an aggregate for the concrete crushed stone of the proper size and color. This can be finished in any way, and the surface thus prepared is permanent, will not fade, deteriorate, scale, nor require renewing. If the desired

Dry Material used	Weight of Dry Coloring Matter to 100 lbs. Cement				Cost of Coloring Matter, ¢ per lb.
	½ lb	1 lb	2 lb	4 lb	
Lampblack	Light slate	Light gray	Blue-gray	Dark slate	15
Prussian blue	Light green slate	Light blue slate	Blue slate	Bright blue slate	50
Yellow ocher	Light green			Light buff	3
Burnt umber	Light pinkish slate	Pinkish slate	Lavender pink	Chocolate	10
Red iron ore	Pinkish slate	Dull pink	Terra cotta	Light brick red	2.5

color cannot be acquired by the use of colored aggregates alone, the cement itself can be given almost any color by mixing it with certain coloring matters. In "Cement and Concrete," by L. C. Sabin, the above color table is given.

The results obtained from coloring are as yet not definite. Some colors seem to fade when exposed to the weather, especially lampblack and Prussian blue. The iron ore seems to be the most permanent in color, but even it gets lighter with exposure.

Stucco

Stucco has been in use to a greater or less extent for ages. The Greeks and Romans were experts in its application, and some of the finest examples of fresco and inlaid tile work are to-day preserved in stuccoed surfaces. In European countries stucco or plastered houses are more common than frame, probably because lumber is comparatively scarce, and because the appreciation of the beauty of these surfaces is greater than here.

Stucco has been employed more extensively for building purposes in warm, dry climates than elsewhere, because in the past only mud or lime plasters have been used. This could not endure in a damp climate, nor could it withstand the sudden and wide changes in temperature of colder climates. The use of cement, however, has changed all this, and the architectural beauty which can be developed in the use of stucco can be enjoyed anywhere.

Stucco to-day is being employed for two purposes — to form the exterior wall of a building, or to renovate an existing structure by giving it a more pleasing appearance. Many old buildings have been made presentable and attractive by putting a coat of stucco over the stone, brick, or wood of which they have been built, forming a permanent finish.

Constituents. — The mixture commonly used in stucco work consists of cement and sand, with the addition of about one part of hydrated lime to ten parts of cement. The cement and lime should be thoroughly dry mixed first, then double the quantity of clean sand added, and the whole mass mixed until it shows a uniform color. Water should then be added until the mixture has a consistency of a stiff plaster.

Method of Application. — As in all concrete work, careful workmanship is an essential of success. When cement plaster is applied to a surface which absorbs moisture, proper adhesion cannot take place; or if a freshly plastered surface is exposed to the heat of the sun, evaporation will take the water from the cement and prevent its proper hardening. To avoid this latter contingency, the surface must be protected with canvas, or frequently sprayed. Stucco should never be applied when the temperature is below freezing, for the water will of course turn to ice and the cement will not harden. Another precaution is to never disturb the stucco after the cement has begun to set.

On Frame Buildings. — Some distinct advantages of a stucco exterior over shingle or clapboard work are responsible for a great and growing popularity of the former. Stucco, when properly applied, is permanently enduring, improves in appearance with age, and has no maintenance charge for painting or renovating. It is claimed that stucco makes a frame house warmer in winter and cooler in summer, and it naturally adds to the fireproof quality of the structure.

Figure 16 shows the usual type of construction employed in stucco work. The framework, consisting of the studs and sheathing, must be made as stiff as possible, for any swaying of the structure will necessarily crack the stucco. The sheathing is well nailed to the outside of the studs; then follows a layer of good, heavy building paper, held in place by $\frac{5}{8}'' \times 1''$ furring strips, placed vertically 9 inches apart. On these strips is fastened either the metal or wood lath, to which the stucco is applied.

To use metal or wood lath is a mooted question. Some reputable architects deplore the use of either, and will not specify the application of stucco to anything but a hollow

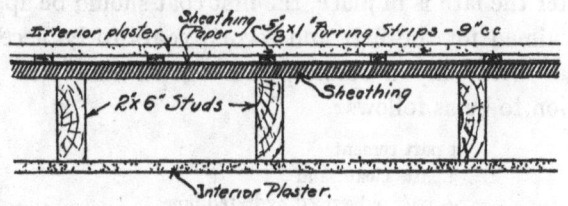

FIG. 16. — Stucco on frame structure. Modern method of application.

tile with a specially shaped side. Other authorities advocate the use of a narrow wood lath, claiming the expansion and contraction to be so slight that no cracks in the stucco result. Metal lath are recommended because there is no absorption of any moisture required by the mortar, the fire risk is decreased, and there can be no cracks to harbor vermin. One essential is that no water get behind the stucco. To prevent this, all roof guttering and downspouting should be put up before the plastering is done. Wood window and door sills should project well from the face of the plaster, and should have a good drip, either by a downward slant, or by a groove rebated in the under side of the sill near enough to its edge that it will not be covered by plaster.

If metal lath is used, it should be properly protected from

corrosion. This can be accomplished in several ways. An expensive but effectual method is to plaster the lath on both sides; but if this is impracticable, the lath may be dipped in a paint made of equal parts of neat cement and water. Immediately after the dipping, the lath should be attached to the furring strips, and the stucco should be applied as soon as the cement has hardened on the metal. A bitumen paint, to which cement will adhere, can be used, but two dippings will be necessary, and the paint should dry for twenty-four hours.

After the lath is in place, the first coat should be applied. It is aimed for the first and second coats to be a cement mortar with only a small percentage of lime, the composition to be as follows:

>   1 part cement
>   2 parts clean sand
> $\frac{1}{10}$ part pulverized hydrated lime

All materials are to be measured by volume. They should be thoroughly mixed dry, and then water is added until the mortar is of the proper consistency for plastering. For the first coat, add one pound of hair to each bag of cement.

In doing the work, the plastering should be started at the top, and carried downward continuously without allowing the plastering to dry at the raw lower edge. If the wall is so wide as to make it impossible to work the full width at one time, make the break at some natural division, such as a door or window. The plaster must be forced through the meshes so as to form a good key; a small-sized mesh, not larger than $\frac{3}{8}''$ by $\frac{5}{8}''$, is preferable, since it will prevent the waste caused by dropping cement through the larger meshes. The thickness of the first coat should be about half an inch: while this coat is still wet. it should be scratched deeply over the entire surface, and then as soon as it can support the second coat, the latter is applied, from $\frac{1}{2}''$ to $\frac{3}{4}''$ thick. This should also be scratched to provide a rough surface for the finish coat. The finish coat should contain no lime nor hair, but should have some reliable commercial waterproofing mixed with it according to the directions given by the manufacturer.

When half-timbering is used, the boards should be rebated as shown in Figure 16, in order that moisture be kept out as much as possible.

Surface Finishing. — Stucco admits of wide variation in surface finish, and almost any effect may be obtained. A few methods are listed herewith.

Smooth. — This finish can be secured by troweling the final coat to an even surface.

Roughcast. — By using plasterer's trowels covered with carpet or burlap a rough-coat finish may be obtained. The irregularity of the surface may be varied by using coarse-grained sand.

Slapdash. — It requires an expert to do this finish well; the method is to throw on the final coat with a paddle.

Pebble Dash. — Apply the final coat rather wet, then throw clean pebbles about $\frac{1}{2}$ inch in diameter into it. Start the work at the top, and throw the pebbles on with a sweeping motion, using enough force to imbed them securely. Care must be taken not to disturb the cement after it has started to set, and in order to avoid this the surface must be covered with the pebbles immediately after the fresh plaster is applied. It is well to have a separate workman handle the pebbles if the surface is of any size; but if it is cut up into smaller separate portions, one of these may be plastered with the final coat and covered with pebbles immediately afterward by the same workman.

In addition to the various finishes that can be given to stucco, it can be colored almost any shade desired. Very beautiful effects can be obtained by properly arranging the various colors at different places in the structure.

Concrete Blocks

The manufacture of concrete blocks has assumed great importance as an industry of recent years, on account of the simplicity of structures built of concrete blocks, and the ease of handling them, the use of forms being obviated. Manufacturers of concrete blocks claim that they excel stone in texture, color, appearance, and durability, when properly made, and are considerably lower in cost.

Any discussion of all block machines is futile, from the fact that there are hundreds of machines upon the market, all differing, some widely, some only in details. The immense variety is an excellent witness both to the inventive genius displayed by concrete men, and to the widespread, sincere interest in block manufacture. There are at least a dozen types of machines — molding face-down, face-up, or side-face blocks; with horizontal or vertical cores; with single, double, or staggered air spaces; using dry, medium, or wet mixtures; and so on, almost indefinitely.

Size. — The best size to construct a block is a question which is best settled by a consideration of the work in hand. A fair average would be, perhaps, one 16 to 24 inches long, and 8 inches high, with a depth varying from 8 to 12 inches. An 8-inch block is amply strong for small residences, but the building ordinances of some cities require a 12-inch block.

Types. — Two general types of blocks are made, faced and unit in construction. A faced block consists practically of two layers, the heavier body of the block being composed possibly of a 1:4 mixture of cement and gravel, while the face is made of a 1:2 mixture of cement and sand. A successful block must have a perfect bond between these two layers, otherwise the face will probably check or crack off, presenting a very bad appearance. The facing mixture ordinarily contains a small percentage of hydrated lime, in order to secure an attractive texture and finish for the surface. Waterproofing may also be included in the facing mixture; in fact, it is desirable. Coloring matter, too, may be incorporated, and blocks of any color may be produced.

In making the unit block, the same mixture is used throughout. Contrary to the usual idea, a comparatively coarse aggregate may still present a pleasing exterior, and the unit block is consequently gaining favor and prominence. The block may be waterproofed throughout, though if a rich enough mixture is used, the block should be sufficiently impervious without waterproofing.

Design of Block Faces. — The selection of a design for the face of a block is so much a matter of personal taste that it may seem useless to attempt to lay down any rules on this subject. The favorite block seems to be one with a flat face, either flat or beveled corners, and with sufficient diversity in size of face to do away with monotony.

Curing. — Irrespective of type, design, color, or face, the block must be properly cured. Every possible precaution must be taken to prevent the drying out of the block during the initial set and early hardening. They must be protected from wind, sun, dry heat, and freezing until they have fully solidified. Two weeks is not too long a time to accomplish this. Even with this, blocks should not be used, except under special conditions, until they are at least six weeks old. A 24-inch block will shrink about

$\frac{1}{16}$ of an inch in that length of time, and if green blocks are placed in a wall the shrinkage will be perceptible.

Laying Blocks. — The best mortar to use in laying concrete blocks is one composed of:

> 1 part cement
> 3 parts hydrated lime, and enough sand to make a rich mortar

The blocks should be wetted thoroughly before laying to prevent the absorption of moisture from the mortar. The mortar joint should be rather thin, from $\frac{1}{4}$ to $\frac{3}{8}$ inch in thickness.

Special Shapes. — Special shapes and sizes of blocks are necessary for the construction of silos, porches, cornices, sills, and other architectural details. The equipment of a good machine includes sufficient forms to make almost any shape of block desired.

Reënforced Concrete

Reënforced concrete is ordinary concrete in which iron or steel rods or wire is imbedded. Reënforcement is required when the concrete is liable to be pulled or bent, as in floors, beams, posts, walls, or tanks, because, while concrete is as strong as stone masonry, neither of these materials has nearly so much strength in tension as in compression. Moreover, concrete alone, like any natural stone, is brittle, but by imbedding in it steel rods or other reënforcements, the cement adheres, and the metal binds the particles together, and the reënforced concrete is then better able to withstand jar and impact.

The idea of reënforcing concrete may be gathered from the following, using Figure 17 for illustration:

Suppose a rectangular, concrete beam be supported at B and at C, with a load applied at A. The beam will be divided into two parts by a horizontal plane, shown in the figure by the dotted line. That part of the beam at A below the dotted line will be in tension; that is, the action of the force A will tend to tear the particles apart. Above the dotted line, that part of the beam at A will be in compression, or will be resisting a tendency toward crushing at that place. At the dotted line itself, there will be no strain, consequently it is known as the neutral axis.

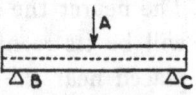

FIG. 17. — Principle of reenforcing concrete.

Now concrete is from six to ten times as strong in compression as it is in tension, and unless the lower part of the beam is treated in some way so as to bring the resistance to the tension in the lower part as high as the resistance in compression in the upper part, the full efficiency of the beam is not attained. By imbedding steel rods, a material very high in tensile strength, in the lower part of the beam, the concrete materials are more firmly bound together, and this, added to the strength of the steel itself, greatly augments the resistance to a tensile strain in this part of the beam. The beam is properly reënforced when there is enough steel in the lower part to increase the tensile strength sufficiently to equalize the compressive strength in the upper part.

Since concrete is a brittle material, and steel a comparatively ductile one, it might be imagined that the stretching of the tension part of the beam would result in the formation of cracks on this surface, leaving the steel to resist all the pull. This has been proved to be true to a certain extent, and it might be supposed that these cracks would admit moisture, resulting in corrosion of the steel. However, while these cracks do reduce the strength of the concrete, they are so minute, and so uniformly distributed, that the reënforcing metal is protected even up to its elastic limit.

Not only must the steel be correctly located, but it is essential to have the proper quantity of metal in the beam. The nearer the steel is placed to the neutral axis, the less will be its reënforcing effect; consequently, it should be placed near the surface in the tension section, but not so near as to cause any cracking off of exterior layers. If the amount of metal is too small, weakness will show itself as soon as the metal reaches its yield point; while if the cross section of the metal is too large in comparison with the area of concrete in compression, the beam, in case of failure, will give way by compression in the concrete. The area of the reënforcing metal in rectangular beams and slabs varies according to conditions from about $\frac{1}{2}$ per cent to $1\frac{1}{2}$ per cent of the area of the cross section of the reënforced beam above the steel.

The actual design of a concrete beam or slab to obtain the highest efficiency of both the concrete and the steel is a very technical problem, and will not be taken up here. It should be intrusted to a competent engineer, who is familiar with the character of the member, and the strength and elasticity of the concrete and the steel.

Various shapes and sizes of steel bars are used for reënforcing; most of them are manufactured so as to make the surface of the rod as irregular as possible, to overcome any tendency of the steel to slip through the concrete, and to give a better gripping surface.

The reënforcing of silos will be considered more in detail in a subsequent chapter devoted exclusively to silos and the methods employed in their construction.

The Strength of Concrete

Concrete possesses its chief value as a building material as a result of its great compression strength. With reënforcing it becomes an ideal material for columns, beams, and floors. Tests of concrete are made to determine either the tensile, compressive, or transverse strength, and from the results of these tests deductions are made as to the comparative strength with other materials.

Plain concrete, that is, concrete without any reënforcing, varies in strength according to:

1. The quality of the cement.
2. Texture of the aggregate.
3. Quantity of cement in a unit volume of concrete.
4. Tensity of the concrete.

The actual strength of concrete in compression, because of the limited capacity of testing machines, can be determined only by experiments upon conparatively small specimens. The actual strength of a good concrete, carefully made and laid, is in all probability somewhat higher than the results of experiments indicate, because specimen blocks cannot contain such a homogeneous mixture as would exist in actual practice. Taylor and Thompson have evolved a simple formula for the determination of the strength of plain concrete, which gives sufficient accuracy for comparing the compressive strength of mixtures of the same materials in different proportions. The formula follows:

Let
P = unit of compressive strength of concrete.
C = absolute volume of cement in a unit volume of concrete.
S = absolute volume of sand in a unit volume of concrete.
g = absolute volume of stone in a unit volume of concrete.
M = a coefficient, varying only with the age of the concrete.

Average values of M are as follows:

AGE	VALUE OF M
7 days	9,500
1 month	12,500
3 months	15,600
6 months	16,900
1 year	18,000

Then
$$P = M\left(\frac{C}{1 + C - (S + g)} - 0.1\right).$$

TABLE OF COMPRESSIVE STRENGTH OF CONCRETE

PROPORTIONS	AGE SIX MONTHS. 40 % VOIDS. LBS./SQ. IN.
1 : 1½ : 3	2720
1 : 2 : 4	2410
1 : 2½ : 5	2130
1 : 3 : 6	1910
1 : 4 : 8	1530

Transverse Strength of Concrete. — The strength of a plain concrete beam is limited by the tensile strength of concrete at the place of greatest strain, which with vertical loading is at the lower surface.

TABLE FOR TRANSVERSE STRENGTH OF PLAIN CONCRETE

PROPORTIONS	STRENGTH IN LBS./SQ. IN.
1 : 2 : 4	440
1 : 3 : 6	226
1 : 4 : 8	157

Mortar

Mortar is composed of lime or cement and clean sand, with just enough water to make a plastic mass. The proportion of sand depends upon the character of the lime or cement.

BUILDING MATERIALS

Cement Mortar. — In mixing cement mortar the cement and sand are first mixed thoroughly dry, the water then added and the whole worked to a uniformly plastic condition. The quality of the mortar is governed largely by the thoroughness of the mixing, the object to be attained being so completely mixing the materials that no two adjacent grains shall be without an intervening film of cement. The chief faults in mixing mortar are not mixing the materials thoroughly when dry, and adding an excess of water in order to facilitate the labor of mixing. An overdose of water is better than an insufficiency, however, for cement is very absorptive.

In mixing by hand a platform or box is essential; the sand should be spread in an even layer, then covered with the proper amount of cement, after which both should be turned and mixed with shovels until a thorough incorporation is effected. The dry mixture should then be piled in a heap, with a crater at the top, and all the water required poured into it. The material on the outside of the crater should be thrown in until the water is taken up, and then worked in a plastic condition.

In order to secure good mixing, it is customary to specify the mixture to be turned a specified number of times with shovels, both dry and wet. The mixing with the shovels should be performed quickly and energetically.

The proportion of cement to sand varies with the nature of the work and the necessity for strength or imperviousness of the mortar. The sand for mortar must be clean, that is, free from loam, mud, or organic matter, sharp and fairly coarse, and not too uniform in size. The water should be fresh and clean, free from mud and vegetable matter. The quantity of water can be determined only by experience, since the nature of the sand and the cement, and the proportions of each, govern it so largely. Fine sand requires more water than coarse to give the same consistency. Dry sand will absorb more water than moist, and a sand composed of porous materials will require more than one composed of hard material.

The purpose for which the mortar is to be used also affects the amount of water used. The consistency of mortar for masonry is such that it will stand in a pile, and not be fluid enough to flow. Mortar for plastering is more plastic.

CEMENT AND SAND REQUIRED FOR ONE CUBIC YARD OF MORTAR

PARTS OF CEMENT: SAND	CEMENT, BBLS.	SAND, CU. YD.
1:1	4.00	0.60
1:2	2.75	0.80
1:3	2.00	0.85
1:4	1.50	0.90
1:5	1.25	0.93
1:6	1.00	0.95

As to the amount of water, it has been found by numerous experiments that, as a general rule, one part of water to three parts of cement by measure, or three and one half parts of cement by volume, is the best, both in regard to convenience in mixing and in the ultimate strength and durability of the mortar.

AMOUNT OF MORTAR REQUIRED FOR A CUBIC YARD OF MASONRY

KIND OF MASONRY	MORTAR, CU. YD.
Ashlar, 18″ courses, 1″ joints	0.035
Ashlar, 12″ courses, 1″ joints	0.075
Brick, standard size, ¼″ joints	0.10 – 0.150
Brick, standard size, ⅜″ joints	0.25 – 0.350
Rubble, small rough stones	0.33 – 0.400
Rubble, large, hammer dressed	0.200 – 0.300

PAINT

A *paint* is a liquid coating applied to wood, steel, iron, or other material for the purpose of ornamentation or protection, or both. It consists of a base (usually a metallic oxide), a vehicle, and a solvent. The vehicle is the liquid part of the paint; in most paints it is either raw or boiled linseed oil, sometimes with the addition of a little turpentine. In enamel paints the vehicle is varnish; in calcimine and other cold-water paints it is a solution of glue, casein, albumen, or some other cementing material, which is sometimes called a binder.

Bases. — The base for most common paints is either white lead or zinc oxide; these, unchanged, form the base of most white paints, while for colored paints various pigments are mixed with them. White lead is a hydrated carbonate of lead, obtained by pouring carbonic acid gas over a mixture of oxide of lead (litharge) and water with about 1 per cent of acetate of lead. It is insoluble in water, but easily soluble in nitric acid, and dissolves when heated, first turning yellow in color. Its chief adulterants are gypsum, whiting or chalk, zinc oxide, and sulphates of baryta and lead. Oxide of zinc is produced by distilling metallic zinc in retorts under a current of air; it will dissolve in hydrochloric acid.

From oxide of lead is produced the red oxide, or red lead, much used as a base for bright red paints. It is manufactured by raising oxide of lead to a very high temperature, just short of fusion, during which it absorbs oxygen from the air, and is converted into the red oxide. By using carbonate of lead and properly regulating the temperature, an orange base, called orange lead, is obtained. Red lead is adulterated with various metallic oxides, with red oxides of iron. and with brick dust.

Iron oxide is produced from the brown hematite iron ores by roasting, separating the impurities, and then grinding. Shades varying from yellowish brown to black may be obtained by altering the temperatures under which it is roasted.

Sulphide of antimony, or antimony vermilion, is a dull orange-red base produced from antimony ore.

The base for most yellow paints is chromate of lead, or chrome-yellow; green is chrome green, a mixture of chrome yellow and Prussian blue. Ultramarine or Prussian blue is the base for common blue paints. Coal tar gives bases of brilliant red, violet, and purple. Most black paints have for a base carbon, either in the form of lamp-black, boneblack, or graphite.

Vehicles. — Linseed oil is the most widely used of vehicles. It is produced by compressing flaxseed. The oil is allowed to settle until it can be drawn off clear. Good raw linseed oil should be pale in color, transparent, and almost free from odor. It improves with age; its drying quality and color may be improved by adding a pound of white lead to each gallon of oil and letting it settle for a week, when the oil is drawn off. The white lead remaining can be used as a base for coarse paint.

Boiled linseed oil is prepared by heating raw oil; it is thicker and darker in color than raw oil, and is not as suitable for delicate work. However, it dries in about one fourth the time required for raw oil, and is valuable on account of this property.

Linseed oil is subject to various adulterations, as by the addition of hemp, fish, cottonseed or mineral oils, which are difficult to detect. Various substitutes for linseed oil, such as fish oil or cottonseed oil treated with benzine, are on the market, as well as numerous patented preparations, under which class comes Japan oil.

BUILDING MATERIALS 57

Solvent. — About the only solvent used in paint manufacture is spirits of turpentine, a volatile oil obtained by the distillation of turpentine from the yellow pine trees of the southern states. The residuum left after distillation is called *rosin*, to distinguish it from the finer resins used in varnish manufacture. Good turpentine is colorless and has a pleasant pungent odor. It is often adulterated with mineral oils, and benzine, naphtha, etc., are often employed as a substitute for it.

Driers. — These are compounds of lead and manganese, dissolved in oil, and thinned with turpentine or benzine. They act as carriers of oxygen between the air and the oil, and their addition makes the paint dry more rapidly. Not more than 10 per cent by volume of drier should be added, since any excess will lower the durability of the paint.

Pigments. — These are added to regular paints of a basic color to obtain other colors. The principal ones are as follows:

Blacks — lampblack, vegetable black, ivory black, boneblack.
Blues — Prussian blue, blue lead, cobalt blue.
Browns — raw umber, burnt umber, burnt sienna.
Reds — red lead, vermilion, Indiana, Chinese, and Venetian red.
Greens — arsenites of copper, cobalt, ferrous oxide of iron, mixtures of blue and yellow pigments.
Yellows — chrome yellow, Naples yellow, yellow ocher, raw sienna.

General Composition

The composition of paint varies with the purpose for which it is to be used and the surface it is intended to cover. If the paint is to be subsequently varnished, it must contain a minimum of oil. If it is to be exposed to the sun, turpentine must be added to prevent blistering; it is also necessary to make paint adhere to old painted surfaces. On new work, the first coat is called the primer, and is chiefly oil, made by adding a gallon of raw linseed oil to each gallon of ordinary paint. Knots and resinous places should be covered with a shellac varnish before oil paint is applied as a priming coat.

Exterior Painting. — For new exterior work, at least three coats are necessary for a satisfactory paint surface. The first, or priming, coat is largely absorbed by the wood. Residences are usually painted with a white lead base, which is sold as a paste containing 10 per cent of oil. White zinc is also an important base. Each has its defects, the white lead having a tendency to powder, and the white zinc becoming hard and scaly; by mixing the two together in the proportions of $\frac{1}{3}$ white zinc to $\frac{2}{3}$ white lead, a product is formed superior to each of its components.

Painting may be facilitated if the trim is painted first, leaving the body color to be laid on neatly against it. The paint should be brushed on with the grain, and each coat should be allowed a week in which to harden before the succeeding coats are applied. The priming coat will require about a gallon of paint for each 300 square feet of surface, the second and third coats being much thinner, a gallon of paint covering about 500 or 600 square feet. The paint for roofs should contain a large proportion of oil, and little or no drier.

The treatment of shingles may result in especially beautiful effects if properly done. Special shingle stains of almost every conceivable color and tints and shades of color are made, which consist of a pigment suspended in creosote or some similar liquid, the creosote having a definite preserving effect. Objection is sometimes made to the odor of the creosote, but this soon passes away; should

BUILDING MATERIALS 59

the rain water collected from the roofs be used for household purposes, it is better that it be diverted from the cistern for a time, until two or three good rains have washed the roof. Creosote is not poisonous, but it is more or less disagreeable in odor.

Interior Painting. — Doors and window frames are given a priming coat before they leave the mill, the priming being omitted on those surfaces which will later be varnished or stained. As mentioned before, all resinous knots should be shellacked before any paint is applied. Following the priming coat should come the puttying, which is done more satisfactorily with a wooden spatula than with a steel putty knife, which cannot be used without marring the surface. The paint for the second coat should have a vehicle which is half turpentine so that it will dry with a dull, or "flat" surface, to which the next coat will adhere readily. The third coat is usually the final one, and may be an ordinary paint, drying with a gloss that may be removed by a light rubbing with pumice stone and water.

Enamel paint, a harder and more expensive paint than oil paint, is made with varnish as a vehicle. It is commonly applied over oil paint which has been slightly roughened with sandpaper when quite dry. When the first enamel coat has hardened, it should be sandpapered or cut with curled hair, and then covered with the final coat, which may be left glossy or rubbed flat as desired.

Varnish. — Varnishes are of two kinds, spirit varnishes, made by dissolving a resin in a volatile oil, of which type shellac is a familiar example, and oil varnishes, in which the resin is mixed with linseed oil and this compound dissolved in turpentine or benzine.

The gums principally used in making oil varnishes are

amber, anime and copal, the last of which is used the most extensively. It is not as durable as amber, and not so expensive. Coach varnish is made from the paler kinds of this gum. Of the softer gums, mastic, gammar, and resin are dissolved in the best grade of turpentine, and make a light, quick-drying varnish, which, however, is not very tough nor durable. The softest gums, lac, sandarac, etc., are dissolved in alcohol to make a quick-drying varnish harder and more glossy than the turpentine varnishes, but not nearly so durable nor so resistant to exposure.

Applying Varnish. — The wood to be varnished first receives a coat of paste filler, which is strongly rubbed in along the grain with a stiff brush, and which, after a half hour's drying, is rubbed off with burlap or excelsior across the grain. Following this, any necessary puttying is done, and in two days the first coat of varnish is applied; after five days it is cut with curled hair or sandpaper to remove the gloss, so the next coat will adhere well; then two or three coats of varnish five days apart, each coat well rubbed except the last, which may be left glossy, or given a flat tone by rubbing with pumice stone and water.

Floors that are to be varnished should receive the treatment above described, using a shellac varnish, which dries rapidly and does not discolor the wood to any great degree. If the floors are to be waxed, a regular floor wax should be obtained, and after one or two coats of shellac varnish have been applied, then five or six coats of wax should be put on at intervals of a week, each coat being well polished with a weighted floor brush used for the purpose. While waxed floors are undoubtedly handsome in appearance, the difficulty and expense of maintaining them in a first-class condition makes the use of varnish more practicable.

BUILDING MATERIALS

Linoleum, a floor covering which is much used in kitchens and bathrooms, may be kept permanently bright and clean by giving it a couple of coats of shellac varnish each year.

Refinishing Old Work. — Exterior work, if properly executed by good workmen with good materials, should last from five to ten years; it may lose its luster, without deterioration in the body of the paint, in which case the surface need only be cleaned and given a coat of oil, to supply the deficiency caused by evaporation. Repainting may be done over an old surface, if it is still smooth, but if it is rough and scaly, it will have to be scrubbed off with a stiff wire brush, or in extreme cases the old paint may have to be removed with the flaring blast of the painter's torch, which so softens the paint that it may be scraped off while hot.

Interior varnished surfaces may be cleaned with a varnish remover, an expensive and highly inflammable compound of solvent liquids, which penetrate old paint and varnish and soften it so that it may be removed with scrapers or brushes. If the interior has been given originally a covering of first-class varnish, all that may be necessary is a thorough washing with soapsuds, followed when dry by a single coat of varnish. To remove old floor wax, which may have dried and would prevent a uniform appearance upon the application of a fresh coat, a ten per cent solution of sal soda in hot water is used.

GLASS

The ordinary glass used as panes for small windows is called sheet or cylinder glass, from the method of its manufacture; it is first blown into the form of a cylinder, cut axially, and then flattened on a stone or steel plate.

The defects of glass are very noticeable, especially the waviness of sheet glass, which cannot be wholly eliminated. Ordinary window glass is sold by the box whatever may be the size of the panes, the aggregate in square feet of glass being fifty, as nearly as the size of the panes will allow.

Sheet Glass. — Sheet glass, without regard to its quality, is graded according to the thickness, as single strength (SS) or double strength (DS). The latter is supposedly of a uniform thickness of $\frac{1}{8}$ of an inch, while the former may be as thin as $\frac{1}{16}$ inch, though there is a wide variation in thickness in the same piece of either kind of glass. With regard to quality, glass is designated as AA for the best, A for the second, and B for the third grade. The AA glass is supposed to be the best glass that can be made by the cylinder process, but as even this may have flaws in it, it requires very careful observation to distinguish the grade in separate panes of good glass. The B grade is used only in cellar or hot-bed sash, greenhouses, etc.

Regular stock sizes in sheet glass vary by inches from 6 to 16 inches, and above that by even inches up to 20 inches in width and 70 inches in length for double strength, and 34 × 50 inches for single strength. The cost of this glass per square foot increases very rapidly as the size of pane increases.

Plate Glass. — This glass differs from the sheet glass in that it is not blown, but poured out in a molten mass on a flat table, rolled to a fairly even surface, and then ground and polished, so that the thickness of any one piece should be almost exactly uniform. It varies in thickness from $\frac{3}{16}$ to $\frac{5}{16}$ of an inch, and is made in various sizes, even as large as 12 by 16 feet. The coat is determined by the size of the glass. Plate glass weighs about $3\frac{1}{2}$ pounds to the square foot.

BUILDING MATERIALS

Special Kinds of Glass. — *Crown glass* is a superior sheet glass, with a finer surface. Special surfaces are given to plate glass used in doors, transoms, or any place an obscure glass is desired. *Ground glass* in the past was much used for this purpose, but the difficulty of keeping a ground surface clean caused it to be supplanted by the figured surface glass. *Prismatic glass*, made with specially designed corrugated surfaces for diffusing light, is now manufactured by several companies. *Rolled-wire glass*, made by embedding rather small mesh netting in the middle of the glass, and furnished with almost any kind of surface and in almost any size up to 4 by 10 feet, is being widely used in skylight and factory window construction.

NAILS

Nails may be classified according to manufacture as follows:

Wrought nails, forged either by hand or machine; make an excellent clinch without breaking. They are seldom used in connection with woodwork.

Cut nails, cut from a strip of rolled steel of the thickness the nail is to be and a little wider than the nail, to admit of the shaping of a head.

Wire nails, made from a stiff steel wire of the same size as the shank of the nail is to be.

Special nails, such as copper, brass, and composition nails, are made to be used in connection with marine and refrigerator work, and in physical laboratories, to avoid the magnetic effects of iron or steel.

Nails are made in almost any conceivable size and shape to suit every class of work; the principal varieties are listed in the table below. Galvanized nails may be procured, and for fastening shingles, slates, and all kinds of roofing where the durability of the nail sometimes governs the worth of the material which it holds, galvanized nails should be used.

From tests it has been determined that cut nails have a holding power about twice that of wire nails, varying from 123 to 286 pounds for four-penny wire and cut nails, respectively, to 703 and 1593 pounds for twenty-penny nails, in pine wood. The relative holding power of various woods is *about* as follows: white pine 1, yellow pine 1.5, oak 3, elm 2, beech 3.2.

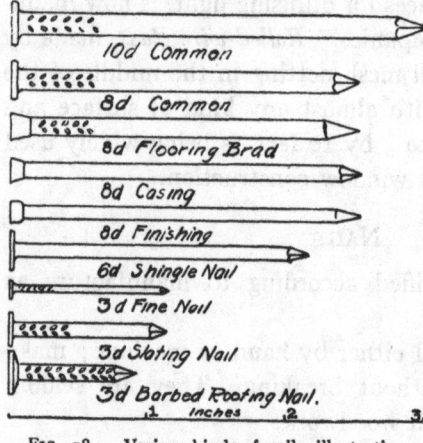

FIG. 18. — Various kinds of nails, illustrating length and special characteristics.

The length of nails is designated by *pennies*, which formerly indicated the pennyweights of metal in the nail. This designation no longer holds good, but the terms are still retained, and the use of them has become so firmly established that it will probably never be changed. The weights run from two to sixty penny, with the corresponding lengths of from one inch to six inches. Common nails have a broad, flat head; casing nails are slightly finer than common nails, and have a head shaped like a truncated cone; finishing nails are still finer, and have a short cylindrical head but slightly larger than the shank of the nail. Spikes are made with diamond or chisel points, and with convex or flat heads.

A good carpenter always uses nails sufficiently large to

securely hold the work, but since a considerable saving can be made by using nails of a size or two smaller, some unscrupulous builders have to be carefully watched. It is well to have the size of nails specified for important work.

For framing, 20d., 40d., or 60d. nails or spikes should be used, according to the size of the timbers. For sheathing, roof boarding, underfloor, and cross bridging, use 10d. common nails. For upper floors of matched flooring, 9d. or 10d. casing nails should be used. Ceiling and partition stuff, when $\frac{3}{4}$ inch thick, is nailed with 8d. casing nails, and with 6d. when of thinner stuff. Inside finish is nailed with finish nails or brads from 8d. down to 2d. in size, according to the thickness of the material. Weather boarding is generally put on with 6d. casing or finish nails; laths should be fastened with 3d. and shingles with 4d. shingle nails, the latter preferably galvanized.

QUANTITY OF NAILS REQUIRED FOR DIFFERENT KINDS OF WORK

1000 shingles — 5 lb. 4d. or 3½ lb. 3d.
1000 lath — 7 lb. 3d.
100 sq. yd. lath — 10 lb. 3d.
1000 sq. ft. weatherboarding — 18 lb. 6d.
1000 sq. ft. sheathing — 20 lb. 8d. or 25 lb. 10d.
1000 sq. ft. flooring — 30 lb. 8d. or 40 lb. 10d.
1000 sq. ft. studding — 15 lb. 10d. or 5 lb. 20d.

WIRE NAIL TABLES

Common Nails & Brads				Spikes			
Size	Length in.	Gauge	No. to 1 lb.	Size	Length in.	Gauge	No. to 1 lb
2d	1	15	876	10d	3	6	41
3d	1½	14	568	12d	3¼	6	38
4d	1¼	12½	316	16d	3½	5	30
5d	1¾	12½	271	20d	4	4	23
6d	2	11½	181	30d	4½	3	17
7d	2¼	11½	161	40d	5	2	13
8d	2½	10¼	106	50d	5½	1	10
9d	2¾	10¼	96	60d	6	1	8
10d	3	9	69		7	0	7
12d	3¼	9	63		8	00	6
16d	3½	8	49		9	00	5
20d	4	6	31		10	⅜	4
30d	4½	5	24		12	⅜	3
40d	5	4	18				
50d	5½	3	14				
60d	6	2	11				

Shingle Nails				Fine Nails			
3d	1½	13	429	2d	1	16½	1350
4d	1½	12	274	3d	1⅛	15	778
5d	1¾	12	235	4d	1½	14	473
6d	2	12	204				
7d	2¼	11	139				
8d	2½	11	125				
9d	2¾	11	114				
10d	3	10	83				

CHAPTER II

LOCATION OF FARM BUILDINGS

IN discussing the location of farm buildings, there are two standpoints that have to be assumed, viz., with reference to the topography of the farm, and with reference to the relation of the buildings to each other. The two are equally important, and the problem of location becomes rather difficult when the subsidiary factors, as convenience, size of farm, prevailing winds, type and use of buildings, etc., are taken into consideration. It is safe to say that in 99 per cent of farms the location of the buildings has been made upon a single secondary consideration, or perhaps two, rather than upon a truly basis one; the site of the building was originally chosen because of its proximity to a spring which may have since been destroyed, or because of a slight eminence which lifted the house out of the miasmic dampness of marshy, low ground, which in the days of modern drainage has become as dry and healthful as any surrounding hill. A square of ground around the dwelling was then fenced off, the barn located just outside of one corner, the corn crib or granary at the opposite, and the smaller buildings, if any, were planted in any place where there was no fence built, or where they would not likely be in the way. There has been no forethought taken in the placing of the buildings whatever.

This state of affairs might be excusable in the case of pioneers to whom these secondary considerations were sometimes of prime importance. It is to be deplored that the same condition exists on farms whose development has been modern, when there was no justification in letting comparatively unimportant things control the whole plan of arrangement. It is evident that no thought has been given to the arrangement of the buildings with relation to each other, or to surrounding conditions: the house has been built with a total disregard of the fine outlook that might have been had from the windows of the rooms most frequented. The barn has been placed with no attempt to screen its undesirable features from the house or the highway; the prevailing winds blow the stable odors directly into the house, and the drainage from the manure flows directly past the gate of the lawn. Many errors are evident in the proper way to approach the house from the highway, and ofttimes there is an absolute disregard of any ornamentation in the way of tree planting — nothing presents itself to view except sharp angles and bare walls of buildings exposed to wind and storm and heat, or there may be a mass of evergreens directly between the house and the highway, obscuring any desirable features the house itself may possess.

This condition is wrong, from any standpoint from which it may be considered. If the owner realized the economic value of the attractive set of buildings on his farm, he would rapidly bring about a rearrangement and remodeling of them to result in the greatest efficiency. The æsthetic value, too, is important; the pleasure to be derived from an attractive and convenient farmstead works subtly and indirectly to increase the actual value of a farm; the farmer's family will certainly be happier and will work more contentedly under conditions inspiring happiness and contentment. Any one can distinguish between an attractive farm, one on which it would be a pleasure to live, and one which is bare and uninviting.

LOCATION OF FARM BUILDINGS

Almost no industry admits of such a wonderful combination of opportunities for the development of health, wealth, and enjoyment of life as does agriculture. While perhaps the economic operation of the farm is of supreme importance in the mind of the farmer, the development of some of the natural beauty peculiar to a farm need not detract a particle from this economic operation, but if properly done, adds many fold thereto.

General Principles of Building Location

To begin with, the home site should be selected so that any part of the farm can be reached without any difficulty or great inconvenience. Many times, in order to avoid small inconveniences, the buildings are located so that part of the fields are more or less inaccessible, or so far away that much time is wasted in going to and from the fields at busy times of the year.

When an approximate location has been decided upon, place the house in the best place available. Try to obtain the most attractive view possible, and build the house so that the view may be advantageously used. The house is by far the most important of farm buildings, though to observe many farms, one would think the exact opposite to be true. At least half of his life the farmer spends in his house, and his wife spends much the greater part of her time there. The farmer's wife is entitled to have a well-built and well-located workshop, in which she manages and contrives to make and keep a happy home, so essential to true success.

If the drainage of the home site is not perfect, this must be attended to, so that good sanitation may be obtained. Plenty of good air and quick drainage of soil are essential.

This can be secured by a location on a fairly dry soil, slightly elevated. Of course, any protection against cold north winds should be taken advantage of, but it is a question whether a windbreak on the west is desirable; cool and refreshing winds should not be deflected during the heated season.

The house should not be located too near the highway, nor is it necessary to have the front of the house toward the highway. Unless because of some special condition the distance between highway and house should not be less than 200 feet, and if the most desirable location for the house be twice or thrice that distance, perhaps so much the better. A park-like entrance drive, the road end of which should be in plain view from the house, should be laid out up to the house-yard gate in a graceful curve; it should be bordered by trees, which should be so arranged as not to interfere with the view. The barn should be located so the prevailing winds will not carry the stable odors toward the house, and the general slope of the land should be from the house toward the barn, rather than the opposite. The barn and any adjacent pens should not be placed in near proximity to the drive, but should preferably be reached by a branch of the main drive. If it is necessary and can be so arranged, another drive should be provided which will not pass near the house, to be used for hauling, etc. The exact position and arrangement of other buildings will be governed by their use; for economy and convenience they should be few and rather compact, though not so close as to increase fire risk. Pens, sheds, and stacks should occupy inconspicuous positions.

LOCATION OF FARM BUILDINGS

Good and Bad Arrangement

In Figure 19 is shown the plan of a farmstead which actually exists. The site is a fairly good one, on moderately level land, with a small eminence to the north on the west side of the highway, and a lower one just south of the farmstead, as shown by the contour lines, which are drawn at intervals of one foot. The house is situated on the slope of the hill, and has a west front, hidden by thick hemlocks; the house is less than fifty feet from the highway, is not close to the pump and milk house nor to the coal shed. The poultry yard is isolated, and the garden is inclosed by a high wood picket fence. There is an entire absence of any sort of an approach to the house, the entrance to the barnyard being at the south end of it, and to reach the house, one must pass the crib, the barn, the machine shed, and go along the feed lot for a distance of almost two hundred feet. This barn lot is partly covered with grass, but near the buildings the grass is worn out by constant driving and tramping. The feed lot, accommodating both hogs and cattle, is adjacent to the house yard.

Let us point out the more prominent bad features of this arrangement. In the first place, the house is much too close to the highway, and is hidden by dense trees; any attractive features it may possess are not taken advantage of; it is too far from the pump and the woodshed and appears entirely isolated from the rest of the buildings. Secondly, the location of the barns and other buildings is particularly unfortunate. They obtrude on the view from the road, are too close to it, and are the most prominent object to be seen along the drive to the house. Finally, absolutely no attempt has been made at any arrangement to improve the natural beauties of the

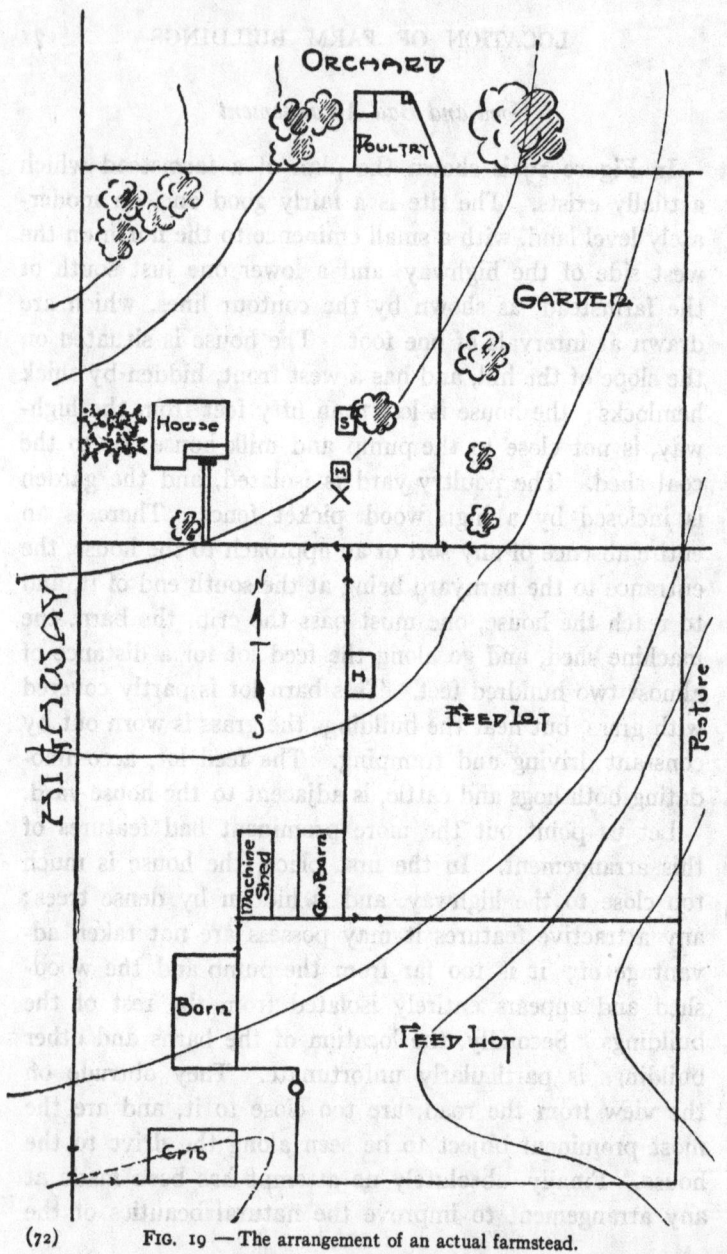

Fig. 19 — The arrangement of an actual farmstead.

LOCATION OF FARM BUILDINGS

site, nor to take advantage of them, resulting in a bareness which is all the more evident in the actual conditions as they exist.

We may assume the main buildings to be in need of replacement, and the relocation of them to use the site most advantageously to be our problem. Beginning with the house, which should properly be the keynote of the arrangement, we place it a little farther down on the slope of the hill, and more than twice as far from the highway as it originally was situated. The windmill, well, and concrete milk house are considered permanent, and this fact precludes putting the house farther from the highway. The hemlocks are removed from the front lawn, as is the fence along the south and east sides of the lawn. A gravel or cinder drive, with its entrance a hundred and fifty feet south of the house, is constructed with a graceful, sweeping, double curve, up to the south front of the house. The other farm buildings are relocated at a greater distance from the highway than before, and with more consideration for economy in time and labor, and with a more definite idea to present a unified whole than was shown originally. A service drive to the barn and crib is put in as a branch of the main drive, and this, as well as the space around the crib, barn, and machine shed, which is likely to be tramped a great deal, is paved with gravel. The poultry yard has been removed to a location between the garden and the feed lot. A hedge of arbor vitæ or osage orange is planted along the farther edge of the service drive from beyond the entrance to the crib, with a gate immediately in front of the barn. A judicious arrangement of tree groups, and a liberal planting of shrub in the right places, completes the arrangement.

What improvements have now been accomplished? To

begin with, the house is more properly located with reference to the highway and appears framed in by the groups of trees at the west side. A broad, unbroken expanse of lawn stretches out to the road. A park-like entrance has been effected, and the drive is carried up to the house in a curve varied enough to prevent monotony. A small porte-cochère may be constructed at the end of the walk leading to the house, and, covered with vines, would add much to the beauty of the arrangement. The barns and other buildings have lost their bold prominence, and have been partially hidden from direct and open view, both from the house and from the highway. Advantage has been taken of some excellent lines of view, especially to the east from the drive immediately in front of the house. The poultry and feed lots have been relegated to positions of relative obscurity though of greater convenience than before. The whole farmstead has become a thing of beauty, comfortable, convenient, and tasteful in arrangement, with a truly homelike atmosphere and appearance. The added value that just this feature gives to the entire farm cannot be estimated.

The particular case which has just been under consideration is but one of the thousands of similar ones which exist everywhere. A little forethought, a little careful planning with the fundamental principles of landscape gardening and of building location well in mind, and a little extra labor, which with its rich returns in the way of æsthetic and material satisfaction should be a labor of love, will transform any barren farm building site into a truly beautiful farmstead.

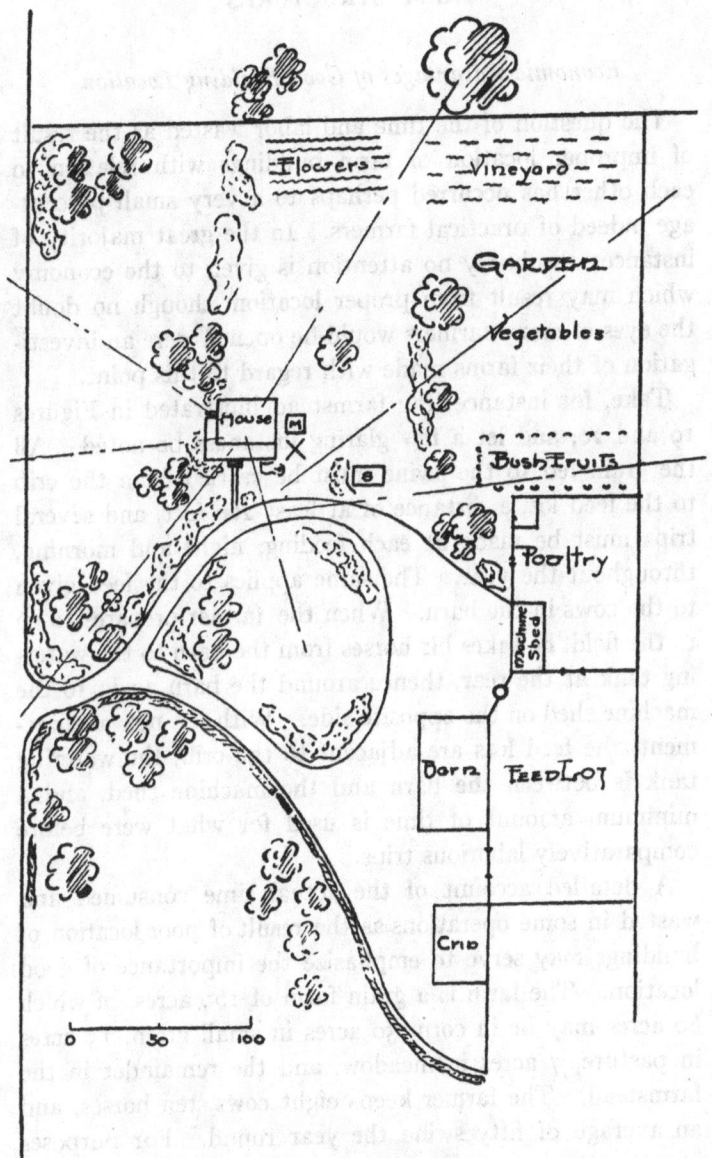

Fig. 20.— Modified arrangement of same farmstead shown in Figure 19 (75)

Economic Advantages of Good Building Location

The question of the time and labor wasted as the result of improper location of farm buildings with relation to each other has occurred perhaps to a very small percentage indeed of practical farmers. In the great majority of instances absolutely no attention is given to the economy which may result from proper location, though no doubt the eyes of many farmers would be opened were an investigation of their farms made with regard to this point.

Take, for instance, the farmstead illustrated in Figures 19 and 20, and let a few glaring instances be noted. All the grain fed to the swine must be carried from the crib to the feed lot, a distance of at least 260 feet, and several trips must be made at each feeding, night and morning, throughout the year. The same applies to the feed given to the cows in the barn. When the farmer prepares to go to the field, he takes his horses from the barn to the watering tank at the rear, thence around the barn again to the machine shed on the opposite side. With the new arrangement, the feed lots are adjacent to the crib, the watering tank is between the barn and the machine shed, and a minimum amount of time is used for what were before comparatively laborious trips.

A detailed account of the actual time consumed and wasted in some operations as the result of poor location of buildings may serve to emphasize the importance of good location. The farm is a grain farm of 160 acres, of which 80 acres may be in corn, 50 acres in small grain, 15 acres in pasture, 7 acres in meadow, and the remainder in the farmstead. The farmer keeps eight cows, ten horses, and an average of fifty swine the year round. For purposes of estimation we may assume a man-hour to be the amount of work done by one man in one hour, and a horse-hour to be the amount of work done by one horse in one hour. The cost of a man hour is estimated at 20 cents, and of a horse-hour at 15 cents.

Amount of grain fed to cows	220 bu.
Amount of grain fed to swine	1000 bu.
Total	1220 bu.
Distance carried	520 feet
Number of trips made	1220
Total distance traversed	120 miles
Distance traversed by a man in 1 hour	2 miles
Total hours consumed	60
Number of man-hours work done	60
Cost of labor in 1 year	$12
Cost of labor in 25 years	$300

Thus we see what an astonishingly large amount of time is consumed in just one small detail. Take another one which is just as bad, and let us see what results.

One corn field of 15 acres is situated in the northwest corner of the eighty on which the farmstead is located. On account of a small stream which is not bridged anywhere along its course through the farm, all trips to and from the field must be made by way of the road, a distance of over 150 rods, or 2500 feet, which could be lessened by 1500 feet were the road made directly to the field through the pasture and across the stream. The farming operations required for managing this field are listed below:

Operation	Time Days	Men	Horses	No. Trips
Plowing	3	1	4	6
Disking (2)	2	1	4	4
Harrowing	½	1	4	1
Planting	1	1	2	2
Rolling	1	1	2	2
Cultivating (3)	5	1	2	10
Husking	7	1	2	14
	7		20	39

This includes only a fair estimate of essential operations, and extra ones, such as hauling fertilizer, cutting stalks, etc., are omitted. Then there is the equivalent of one man making 273 trips and of one horse making 780 trips back and forth from the field over an unnecessary distance of 1500 feet.

Total extra distance man travels	= 409,500 feet = 76 miles
Hours consumed at 2 miles per hour	= 38 hours
Man-hours of work consumed	= 38 hours
Cost of extra man-hours at 20 cents	= $7.60
Total distance horse travels	= 1,170,000 feet = 212 miles
Horse-hours of work consumed	= 106
Cost of extra horse-hours at 15 cents	= $15.90
Total cost of extra work	= $23.50

This is just for one year; assuming a three-year crop rotation, the total loss in thirty years' farming, counting the loss in years in which corn is grown in the field, is $235.

These two examples, chosen at random from the many bad features in the arrangement of the farmstead, show the immense importance of a very careful study of the circumstances and conditions controlling the arrangement of buildings, both with relation to each other and with relation to the farm itself.

In the rearranged plan shown in Figure 20, the economy of the arrangement is shown at a glance. The crib and feeding lots are adjacent; the barn, watering tank, and machine shed are in natural sequence; and should the occasion arise whereby an extra building becomes necessary, it can well be placed east of either the barn or the crib. In fact, a series of buildings could be constructed along a sort of a midway, the beginning of which is shown between the crib and the barn, and the economy of the arrangement be still maintained. This midway is the natural direction of expansion should enlargement of the farmstead ever be found necessary.

CHAPTER III

BUILDING CONSTRUCTION

A THOROUGH knowledge of the details of ordinary building construction is absolutely essential to any one who presumes to plan or superintend any building operations. The prospective architect or superintendent, the latter being often the owner, should be familiar, not only with the kinds, qualities, and grades of lumber in his locality, but with the cost and comparative value of all kinds of building material employed from the foundation up. He should know the names of the various pieces of timber which go into a building, the names of foundation materials and parts, and of the hardware with which the building is equipped.

With the idea in view, then, of enabling the reader to acquire this knowledge in a systematic manner, we shall take an ordinary dwelling house, and follow its construction from the laying of the foundation to the fitting of the interior woodwork.

FOUNDATIONS

The first operation to be employed is the staking out of the foundation. This should be very carefully done, the principal corners being located by a small nail driven into a stake to show the exact intersection of the lines. Six or eight feet from the corner three large strong stakes, 2 × 4, are driven firmly into the ground as shown at *A*, *B*, and *C*, in Figure 21, and braced as shown in Figure 22. The building lines are then marked at *D* and *E* upon these boards, which should be four or six feet long. In this way the building lines are of easy reference until the first story is begun, when the stakes and boards may be removed, since they are no longer necessary. The accuracy of the work may be determined by measuring the diagonals *FG* and *HK*, which should be of the same length.

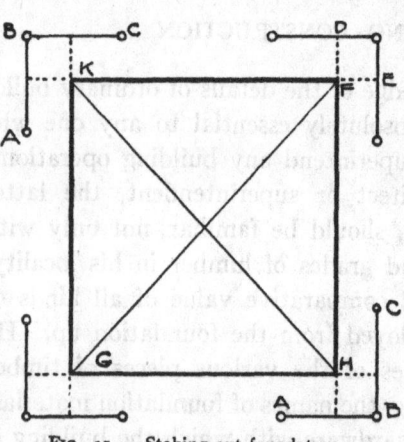

FIG. 21. — Staking out foundation.

After the staking out has been completed, excavators are set to work to remove the earth to the required depth, this being controlled by the height of the first-floor level above the grade line, which may vary from two to five or six feet in ordinary cases, and by the height of the basement story. The latter should never be less than seven feet from the basement floor to the bottom of the first-floor joists; seven and a half or eight feet is much better, and the extra cost is slight. Consideration should be made of the fact that the basement floor itself is at least four inches thick.

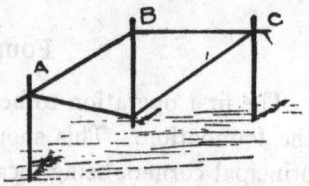

FIG 22 — Stakes and bracing.

The earth removed in excavating should be taken care of for subsequent use in grading the ground immediately surrounding the house. The black earth should be piled

BUILDING CONSTRUCTION 81

separately, and applied on top of the clay when the grading is being done, since a much better lawn can be made on good soil.

No part of a building is more important than the foundation, and most of the cracks and failures in buildings will be found to be the direct result of careless work in building the foundations. In the first place, the foundation should be placed deep enough to afford a firm footing upon comparatively solid earth; secondly, the footing or base of the foundation should be wide enough to adequately support the foundation and superstructure above it.

For ordinary buildings probably the controlling factors in the depth of foundations will be the depth of the basement or the frost line. Even in localities where there is no frost, or in houses where modern heating systems are installed with a furnace in the basement which supplies enough heat to prevent freezing, the foundations should be carried down to a depth of three feet below the surface of the ground, so as to avoid the annoyance of the action of surface water.

Foundations for dwelling houses may be of any one of the following materials: coursed stone, rubble, brick, concrete blocks, or of monolithic concrete. Coursed stone makes an attractive foundation when laid in a darker-colored cement mortar; the same may be said of rubble. Perhaps the most widely

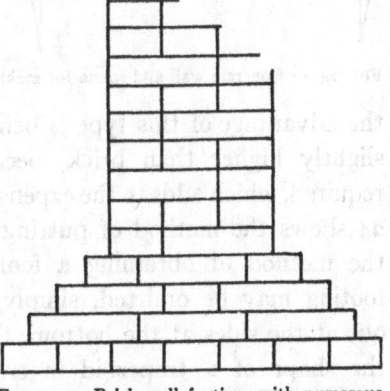

FIG. 23. — Brick wall footing, with numerous header courses.

used foundation material, however, is brick laid in ordinary lime mortar. The construction of common brick foundations is shown in Figure 23. The wall is made about 12½ inches thick, or 1½ times the length of the brick, and is firmly held together by laying numerous courses of headers, as shown, heading in both directions. The footing may be made of almost any width, spreading out the width of a brick at a time until the desired width is reached. Concrete block foundations are built in a similar manner.

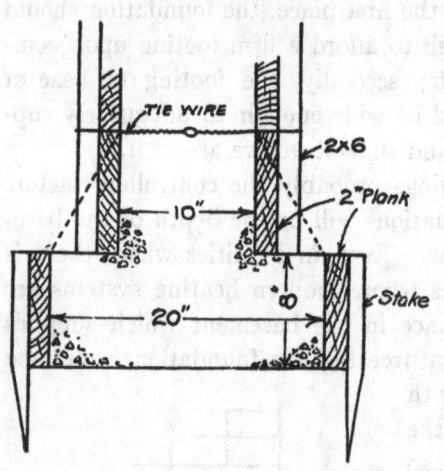

Fig. 24. — Concrete wall and forms for making it.

The use of monolithic concrete foundations, with single or double walls, is becoming more and more prevalent, as the advantage of this type is being perceived. The cost is slightly higher than brick, because generally forms are required, which adds to the expense of construction. Figure 24 shows the method of putting up the forms, as well as the method of obtaining a footing. The forms for the footing may be omitted, simply letting the concrete run out at the sides at the bottom, thus forming a foundation the shape of a trapezoid in cross section. Double-wall construction is rather expensive, since a double set of forms is required, but the air space acts as an excellent insulator, and the sweating of walls is almost absolutely obviated. In any case, where concrete foundations are used, waterproofing should be mixed with the concrete, in order to prevent surface water from finding its way into the basement. The ease with which water oozes through brick masonry sometimes precludes the use of brick foundations, though they may be protected by an exterior coating of tar, asphaltum, or other water-excluding material.

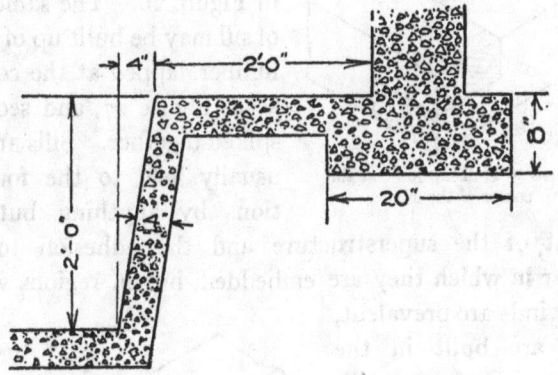

FIG. 25. — Basement shelf. An extension of the concrete footing.

A modification of the ordinary straight foundation wall is shown in Figure 25. The foundation wall itself is not carried down so deep as ordinary, and the inner part of the footing is widened so as to form a very convenient shelf. The bank extending to the floor is covered with a 4-inch thickness of concrete, with a slope of 4 inches from the vertical. The advantage of this modification is very evident, inasmuch as a considerable saving is effected in the amount of material used in the foundation itself, and as the shelf will be a very convenient place to put boxes, jars, etc., when cleaning and scrubbing the basement floor.

FRAMING

The sills which are placed directly on the foundation and which should be firmly bedded in mortar may be of various forms. The simplest sill is made of a 4 × 6, halved together at the corners as shown in Figure 26. The same sort of sill may be built up of 2 × 6 lumber, lapped at the corners as in Figure 27, and securely spiked together. Sills are not usually held to the foundations by anything but the weight of the superstructure and the adhesion to the mortar in which they are embedded, but in regions where high winds are prevalent, bolts are built in the masonry, and the sills are laid with these bolts extending through them, and held in place by nuts screwed on the bolts.

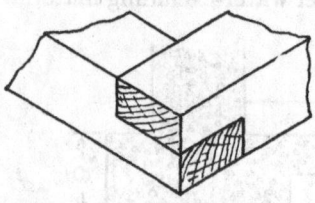

FIG. 26 — Half lap joint at the corner of the sill.

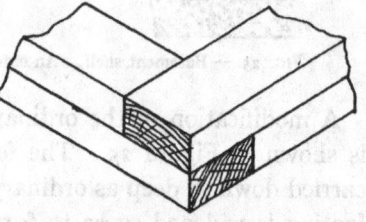

FIG. 27. — Plank sill.

In long buildings, a single length of timber may not be sufficient to form a whole sill, consequently one or more pieces must be joined together, preferably by a beveled half lap, shown in Figure 28. The lasting qualities of sills may be greatly augmented by the application of a protecting coat of good paint, both to the surface of the sill and to its ends and joints.

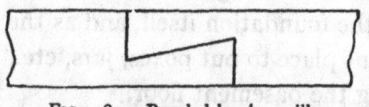

FIG. 28. — Beveled lap in a sill.

A box sill is constructed as shown in Figure 29. It is a rather good form of construction, if given a coat of paint, since it affords a better base to which to fasten the studs than does the plain sill.

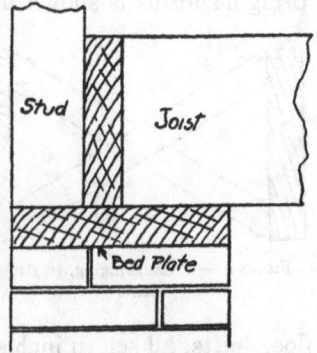

FIG. 29.—One form of a box sill.

Particular care must be exercised to get the sills absolutely at right angles to each other wherever such should be the case, since the appearance of the building depends upon it. This may be done by the 3–4–5 method, laying off a distance of 3 feet in one direction from the corner, and 4 feet in the other direction, and then seeing that the distance between these points is exactly 5 feet.

Joists are those pieces of timber which are laid horizontally on edge upon the sills and which directly support the floor. There are various ways of fastening them; they may be placed immediately upon the top of the sill and securely spiked in place, but unless they are all of the same depth, they will not form level support for the floor. If the under edge of the joist is notched slightly, as in Figure 30, this inaccuracy can be obviated, but the notch should not be cut so deep as to weaken the joist.

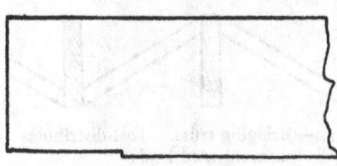

FIG. 30.—Sized joist.

Since the joists are set on edge, they will have a tendency to tip toward one side, especially if there is any warp or twist in them. To reduce this tendency and to strengthen the joists themselves, pieces of 2 × 3 or 2 × 4 are inserted diagonally in the spaces between the joists, as shown in Figure 31. The little truss which a joist and the adjacent bridging forms is shown in Figure 32, illustrating the distribution of concentrated loads. The bridging should be put in at intervals of not more than 8 feet in span. For ordinary small houses the size of joists is 2 × 10 for the first floor, 2 × 8 for the second floor, and 2 × 6 or even 2 × 4 for the attic floor joists, all set 16 inches on center.

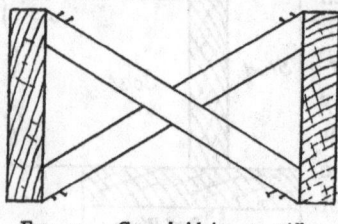

FIG. 31. — Cross bridging, to stiffen joists.

The studding, or upright supports for the wall, may be fixed in several ways, the simplest being where the studding rests upon the sill and is nailed to the joist. Other methods are shown in Figure 29. Studs for small houses usually consist of 2 × 4 stock, set 16 inches on center; the reason for this is that standard wood lath, which are nailed to the stud to hold the plaster, are 4 feet long, and will thus touch four studs. Corner studs are either 4 × 4 or 2 × 4 double.

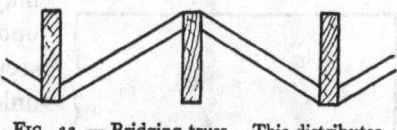

FIG. 32. — Bridging truss. This distributes concentrated loads.

If the framing of the house is to be of the *balloon* type, the studs will extend from the sill up to the plate which supports the rafters, and the second-floor joists will be secured by supporting them on a 1 × 4 or 1 × 6 girt set into the outside edge of the studs, and nailing them to the studs with 20d. nails. In a second type of framing also much used in good construction, the studs will extend only to the second floor to a large horizontal timber, which in turn is mortised into heavy corner posts, a second tier of studs being used for the second floor. This type of framing is called the *braced* frame, from the fact that all sills, posts, girts, and plates are composed of heavy timbers, are all mortised and pinned together, and are braced by 4 × 4 or 4 × 6 braces between each adjacent and vertical and horizontal timber, the braces being mortised and pinned to the timbers they connect. This framing is very strong and substantial when properly done, but is much more expensive than the balloon type described above.

Studding is always doubled around windows and doors, and where the opening is more than 4 feet, the header, or cross timber above the opening, is trussed, as shown in Figure 33. In providing openings for windows and doors of a specified size, the finished dimensions are meant, and in setting the studs, allowance must be made for the width of jambs and frames.

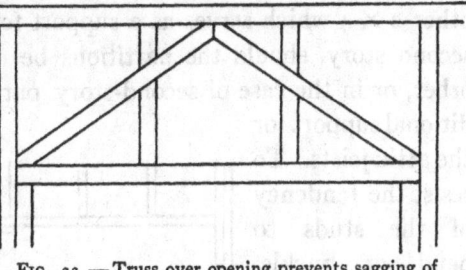

FIG. 33. — Truss over opening prevents sagging of door frame.

Openings for doors should be about 4 inches wider than the door, and since windows are generally designated by certain glass dimension, an allowance of 10 inches above the width of the glass will admit the sash frame and sash-weight pockets. For casement windows only a 6-inch allowance need be made.

To give additional strength to the balloon frame, diagonal braces are set flush into the studs at each corner of the building, these braces being of 1 × 6 stuff.

The interior walls, or partition walls, of a house are built of studs, usually 2 × 4, placed 16 inches on center, leaving such openings as may be necessary for doors, etc., which are trussed and framed similarly to exterior openings.

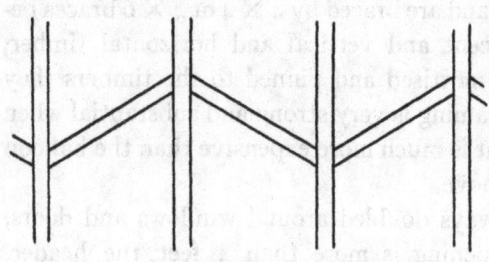

FIG. 34. — Single stud bridging. Prevents studs from buckling.

The support for the studs usually consists of a shoe, a 2 × 4 piece to which the bottoms of the stud are nailed; at the top of the studs is placed another 2 × 4 which serves as a support for the studs of the second story, should the partitions be directly over each other, or in the case of second-story partitions, as an additional support for the attic joists. To resist the tendency of the studs to bend or buckle, single bridging is set in diagonally between them, as shown in Figure 34.

The shrinkage of timber must be given careful consideration in wall framing. Ordinary lumber will shrink from ¼ to 1 inch per foot in width when thoroughly dried, but does not shrink to any appreciable extent in its length. In designing the frame of a house it

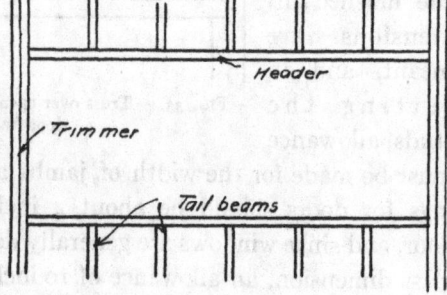

FIG. 35. — Framing floor opening.

BUILDING CONSTRUCTION 89

is of great importance that the members be arranged to permit of equal shrinkage in all parts of the house, otherwise serious difficulty will result. For instance, if the studs and joists be both resting on the sill, one on end, the other on edge, with the baseboard nailed to the stud and the floor to the joist, the latter in drying will draw the floor away from the corner mold perhaps a half inch. Unless the designing has been properly and carefully done, similar discrepancies will result at other places.

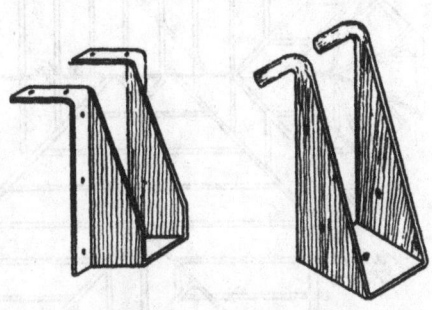

FIG. 36 — Joist hangers.

In cutting holes through the floor, for fireplaces, stair walls, etc., where ends of joists project without any support from above or below, the framing is effected as shown in Figure 35. The tail beams may be fastened to the header by a mortise and tenon joint, the header in turn being mortised into the trimmers, or they may be supported by joist hangers or stirrup irons, shown in Figure 36. The trimmers have to be of extra strength, since they have to support the additional weight incurred by the attachment of the headers. The distance from the inside of any flue to the

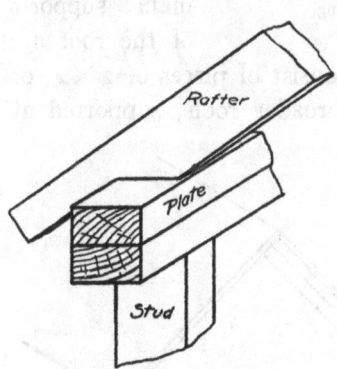

FIG. 37. — Roof framing at plate.

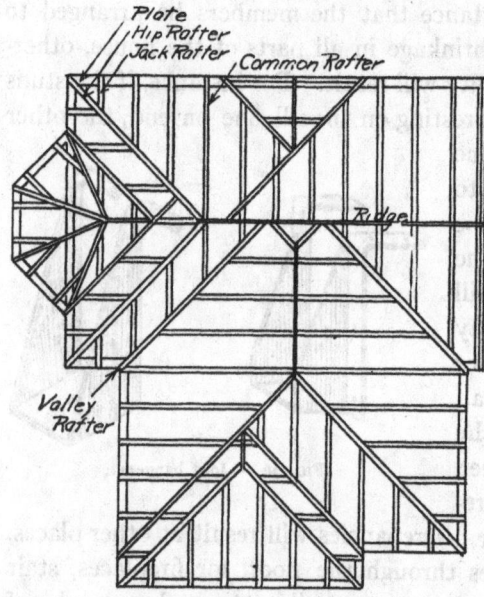

Fig. 38.—Typical roof framing.

face of the joist or header should never be less than 8 inches, nor less than 4 inches from a chimney; if this precaution were always observed, much fire loss would be eliminated.

Roof framing is comparatively simple, but it is often improperly done. The main supports of the roof are called rafters; they usually consist of pieces of 2 × 4, or 2 × 6 in larger houses with broader roofs, supported at the lower end by a plate, two pieces of 2 × 4 nailed firmly to the top of the studs, and held at the upper end either by the opposite rafter or by a hip rafter. The arrangement of the timbers in a typical roof is shown in

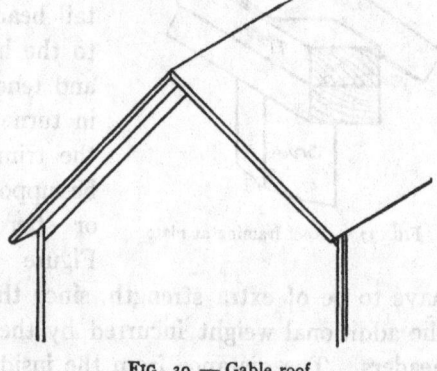

Fig. 39.—Gable roof.

BUILDING CONSTRUCTION 91

Figures 37 and 38. The framing of a roof is sometimes a rather complex problem, especially when the house is not square or rectangular in shape, but has projecting ells of various widths, and with bays and dormers. Figure 38 shows the complete framing plan of the rod of a rather irregular house, with all the component parts of the roof framing indicated.

Houses are usually built with the gable roof, Figure 39, or with the hip roof, Figure 40. The former is simpler in framing, for there are no hip rafters, and valley rafters occur only when dormer windows are put in; however, practically the same amount of roof sheathing and shingles is used in one type as in the other, so the extra cost of lumber in the gable is a disadvantage. The hip roof, too, provides a means of apparently reducing the height of the building, which is sometimes a consideration.

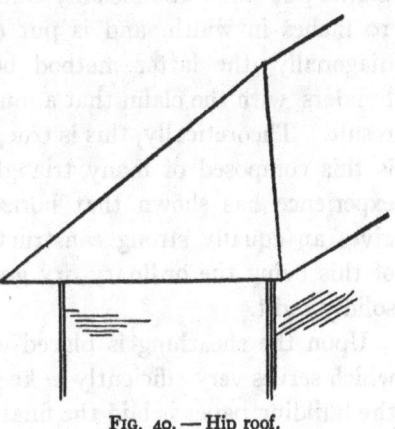

FIG. 40. — Hip roof.

The degree of slope of a roof is indicated by a term known as the *pitch*. The pitch is a fraction in which the *rise* of the roof, or the vertical distance from the plate to the ridge, is the numerator, and the *span*, or the width of the building, is the denominator. Pitch is usually expressed by some simple fraction, as ½, ⅜, ⅓, ¼, etc. To illustrate: for a building 24 feet wide to have a roof of ⅜ pitch, the rise must be 9 feet.

Thus far there has been discussed only the general framing of the house; this must be clearly understood and kept in mind in order that the details of construction may not prove confusing.

Walls

After the studs have been erected and firmly secured and braced, an exterior covering of 1-inch boards, called sheathing, is put on. This usually consists of ship lap, 6, 8, or 10 inches in width, and is put on either horizontally or diagonally, the latter method being preferred by many builders, with the claim that a much stronger structure will result. Theoretically, this is true, since the resulting frame is this composed of many triangles which are rigid; but experience has shown that horizontal sheathing actually gives an equally strong construction, a common example of this being the ordinary dry goods box, which is indeed solidly built.

Upon the sheathing is placed a layer of building paper which serves very efficiently in keeping out cold; then over the building paper is laid the final exterior covering of the wall. This may be weatherboarding, shingles, or stucco, as the fancy of the builder desires. Weatherboarding is made in widths of 3 or 6 inches, and tapers from a thickness of $\frac{5}{8}$ inch at one edge to $\frac{3}{8}$ or $\frac{1}{4}$ inch at the other; it is laid horizontally in shingle style, each successive board overlapping the one below it by 1 inch. This form of wall exterior is very widely used; it requires frequent painting to preserve the wood and to retain an attractive appearance. Shingles are used quite extensively in some parts of the country, and because of their beauty and durability are coming into vogue everywhere. They are economical also, since a poorer grade of shingles can be used on a wall than on a roof, and since the exposure to the weather can be greater on the wall than on the roof, a thousand shingles covering about 150 square feet of wall surface. Shingles are usually stained, instead of painted, with a stain containing creosote, a good wood preservative. Some delightfully attractive effects can be obtained with shingles by using proper treatment, for shingle stains can be obtained in almost every conceivable shade. The use of stucco as an exterior wall surface is common in Europe and Mexico, but has not been generally accepted in the United States until comparatively recently. The difficulty has been experienced that the stucco was not permanent; small cracks, would appear which permitted the ingress of moisture, the moisture causing the stucco to expand and further extend the cracks, until finally the stucco fell off. Modern methods have to a great extent obviated this difficulty, and many fine and expensive houses are being constructed with stucco exteriors. Stucco appeals to many people on account of its soft and pleasing tones, and both when used alone and in connection with half timbering (strips of wood dividing the wall surface into panels), presents a very attractive exterior.

The interior of a wall is generally covered with lath, to which the plaster is applied. Sometimes a pat-

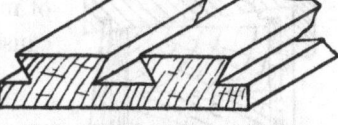

FIG. 41. — Byrkit lath. Used in stucco work.

ented form of lath, known as Byrkit lath, illustrated in Figure 41, is used; since it is practically a solid piece of wood, it serves the purpose of both sheathing and lath, and adds much to the strength and rigidity of the walls. Indeed, it is sometimes used instead of sheathing, the exterior wall covering consisting solely of weatherboarding; this, however, is a cheap and unsatisfactory form of construction, and its use is not advised.

The addition of corner boards, strips of wood 1 inch thick and varying with different houses from 4 to 8 inches in width, sometimes adds to the attractiveness of a house, by adding a border which can be painted the same color as the window and door frames, thus attractively defining the wall contour. Often a "belt" is nailed about the entire house at the top of the first-story or the bottom of the second-story window frame; this may serve to separate one variety of wall covering from another, as shingles from stucco, etc. A baseboard is usually applied at the base of the wall, just above the foundation.

All belts, baseboards, and window frames should be protected by a water table, or cap, as in Figure 45, to prevent the ingress of moisture, which would in time cause rot.

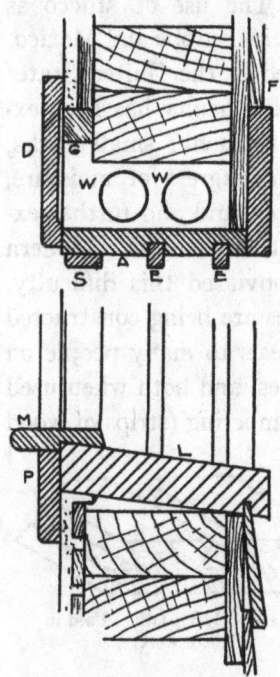

FIG. 42 — Window framing
For an ordinary frame house

WINDOWS

The framing of windows is one of the most important details in house construction. The common method of constructing window frames in wooden buildings is shown in elevation and cross section in Figure 42. Essentially, the frame consists of the pulley stile A, to which are attached the parting

BUILDING CONSTRUCTION 95

strips *EE* and the stop head *S*, the outside casing *B*, the sill *L*, and the headpiece, not shown in the figure, but corresponding at the top to the sill below. The inside casing *D*, the stool or inside sill *M*, and the apron *P*, are considered as parts of the interior trim, and not part of the frame. *WW* are the sash weights which are used to balance the sash and hold it in any desired position. Sash weights are usually of cast iron, but for heavy sash, and for wide and low sash, lead weights are used. In hanging sashes the weights for the upper sash should be about ½ pound heavier than the weight of the sash, and for the lower ones, about ½ pound lighter.

Doors

One thing in building construction that deserves more attention than it usually gets is the matter of setting door frames. The occurrence of a door that does not hang straight, that requires a slam perhaps to latch it, is altogether too common, and while the carpenter may insist that the fault lies entirely with the door, in the fact that it is warped or twisted, in many cases it is due to inaccuracy of the frame itself.

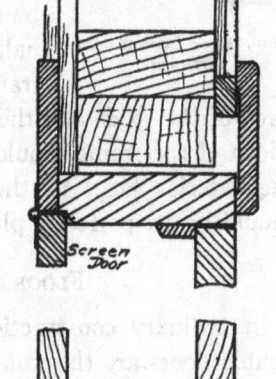
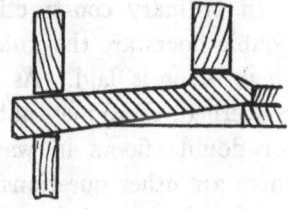

FIG. 43.—Exterior door framing.

In Figure 43 is illustrated the common method of framing for an exterior door. The frames for all outside doors should be made of plank not less than 1¾ inches thick, with the outer edge rebated for a screen or storm door, if desired.

In place of the fashioned sill shown in the figure, a plain plank is sometimes used with a narrow threshold placed under the door. This is sometimes a desirable feature, especially in doors where continued wear on the sill would soon destroy its usefulness, whereas a threshold can easily be replaced when worn.

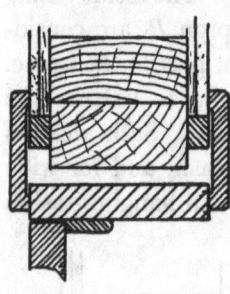

FIG. 44. — Interior door framing.

Interior doors are framed as shown in Figure 44. The studding should be set with a clearance of at least a half-inch to allow of plumbing the frame; wedges are then driven in back of the studs and the frame nailed to the studs. The width of the frame should be exactly the distance between the exterior faces of the grounds, which should be set perfectly plumb.

FLOORS

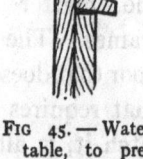

FIG 45. — Water table, to prevent ingress of water.

In ordinary construction of good type, double floors are the rule, though often a single floor is laid. As far as warmth is concerned, there may not be any necessity for double floors in warm countries, but there are other questions to be considered besides the one of heat and cold. During the construction of the building, the rough under floor is a great convenience, almost indispensable. The finish floor is usually of fine wood which would be badly marred if subjected to the rough usage which the subfloor gets during house construction. If anything like good construction is desired, it is false economy to dispense with the subfloor, no matter what the climate; aside from the reasons given above, it adds largely to the strength not only of the floor, but of the whole building as well.

The subfloor, composed of inch stuff, usually ship lap 6 or 8 inches wide, is laid diagonally upon the joists which have been properly sized so as to present a uniformly level upper surface, and is nailed with ordinary 8d. or 10d. nails. The cheapest grade of lumber may be used so long as it is sound and of uniform thickness, in order that there be no inequalities in the upper flooring. The subflooring should be extended to the exterior wall covering between the studs to form a base for a fire stop, which may be of cinder concrete packed to a depth of 6 or 8 inches in each rectangular inclosure formed by a pair of studs and the exterior and interior walls.

The upper or finish floor is laid upon the rough flooring, with or without a layer of building paper or similar substance between the two. Finish floor is made in two thicknesses ordinarily, $\frac{3}{8}$ and $1\frac{3}{8}$ inch, and in a number of widths, from 1 inch to 6 inches. For floors that are to be left uncarpeted the width should not exceed 4 inches; indeed, the $2\frac{1}{2}$-inch width seems to be the most popular. Floors that are to be covered with carpet or linoleum may be of softer wood and may be 6 inches in width, if it is well seasoned, otherwise less, to prevent the formation of ridges which might result from the warping of a rather wide board.

The woods most used in finish floors are oak, plain or quarter-sawed, birch, maple, and hard pine, plain or comb-grain. The hard pine is the cheapest, and is used in kitchens and in buildings where the cost must be kept down. If used where the floor is exposed, only comb-grain stock, corresponding to quarter-sawed in oak, should be employed. Maple flooring is in great favor for floors which are subjected to much heavy wear. For parlor and hall work, and for entire floors in the better class of residences, oak is generally used; quarter-sawed oak is the more attractive, but its cost is much greater.

Finish floor should never be put on until all the plastering has been done in the house and is thoroughly dry; in fact, for the best results, it should be applied the very last thing in the finishing of a house. Most finish flooring is matched, though with a subfloor it is not really essential, and the better qualities are grooved on the underside so as to admit of easier conformation to slight inaccuracies in the subfloor, and to prevent warping.

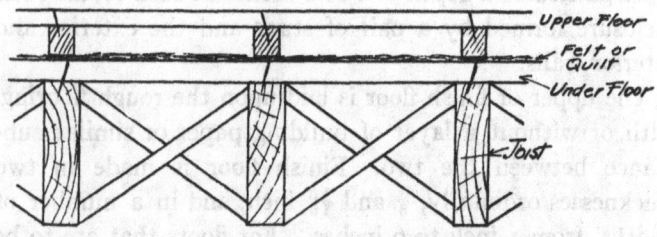

FIG. 46. — Floor deafening. Inexpensive and efficient.

It is quite desirable and almost essential that in dwelling houses the conduction of sound through walls and floors be prevented as much as possible. Usually this is attempted by lining the walls and floors with some sort of material that is expected to absorb the sounds, but this method is not altogether satisfactory, since the lining is not altogether efficacious and there is more or less solid connection between the floor and ceiling in the way of nails or joists. In Figure 46 is illustrated a method of floor deafening which is recommended by Kidder as a most effective procedure in wooden buildings. The efficacy results from the fact that there is no solid connection between the subfloor and finish floor, the cleats for the finish floor being placed directly upon the lining of felt or quilt with no nailing or other fastening. However, this construction is not practicable where only ⅜-inch flooring is used, since to adequately support the floor the cleats would have to be so closely set as to make in effect only a second subfloor, which compresses the lining to such an extent as to partially destroy its deafening value.

The usual materials for deafening are in the nature of paper felts, such as quilt, building paper, asbestos, etc. Mineral wool is also used to some extent, and its fireproof qualities and the nonelastic texture formed by its minute pliant fibers render it quite valuable. A depth of 2 or 3 inches of the wool, upon sheathing lath to which the plaster is applied, will greatly increase the comfort of the room below. Cinder concrete, a weak mixture of cement and cinders, can be applied in the same way, and is cheap and effective.

Roofs

Upon the rafters, which constitute the main support of the roof, is nailed the roof sheathing, inch stuff which varies in character with the type of roof covering. For shingles, sheathing may be of almost any width, though usually 1 × 4 strips are used, placed not more than 2 inches apart, and securely nailed to the rafters. The shingles are then fastened to the sheathing by means of shingle nails, the nails being driven so that they will be covered by the next succeeding tier of shingles above; the shingles themselves are laid with from 4 to 5 inches of their butts, or thick ends, exposed to the weather. The first or lowest tier of shingles is always made double, and at the ridge they are protected by ridge boards, which in turn are protected by a roll of metal which prevents moisture from entering.

Angular places in a roof, such as valleys, around chimneys, etc., are made water-tight by means of metal strips, or flashing, which is arranged so as to conduct all the water on to the roof.

Shingles are perhaps the least durable of the wood material which enters into building construction. They are made of the best woods, and in several grades. Red cedar and cypress are the woods most used for shingles manufacture, the cypress seemingly being the better. Shingles are sold in bunches containing the equivalent of 250 shingles 4 inches in width, though the actual width of the shingles may vary from 2 to 16 inches. Cypress shingles are usually 18 inches long and measure "5 to 2" in thickness, this indicating that the butt thickness of 5 shingles is 2 inches. Cedar shingles are usually 16 in length and butts "6 to 3" in thickness.

If any kind of roof covering is used other than shingles, the roof sheathing is laid close, with no intervening cracks, to provide a better nailing surface. The method of application of other types of roofing has been described in a previous chapter.

The construction of the cornice, or that part of the roof projecting beyond the plate, is a matter which admits of the widest variation in the treatment. A *boxed cornice* is one in which the projecting ends of the rafters are completely inclosed, this term being used in distinction to the *open cornice*, in which the rafters are left exposed. The details of the construction of some simple though effective boxed cornices are shown in Figure 47. On the middle one are indicated the members of the cornice; *a* is called the fascia, *b* the plancher, and *c* the frieze.

The gutters, for collecting the roof water, may be constructed as shown in the first construction, Figure 47, the

BUILDING CONSTRUCTION

trough proper consisting of some sort of metal, either tin, galvanized iron, or copper. It is important that the gutters have the proper slope toward the downspouts, and that the connection between gutters and downspouts is properly protected by a wire screen to keep out leaves, twigs, and the miscellaneous litter that always collects on roofs. The sort of gutter illustrated in the figure has the disadvantage

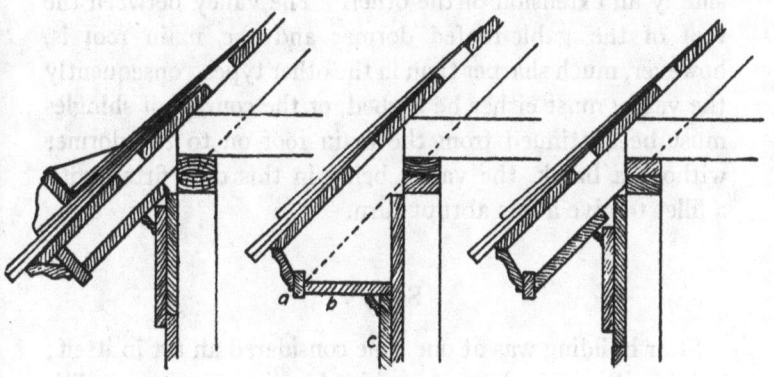

FIG. 47. — Boxed cornices.

of projecting up from the slope of the roof, catching and holding snow and any material which would otherwise slide off; the correctly built gutter is below the edge of the roof, the line of the slope passing above it. The cheapest form of gutter consists of a trough hung from the lower edge of the roof, but this type is extremely liable to be blown down or so damaged by wind as to render it useless.

DORMERS

The term dormer is applied interchangeably either to a vertical window in a roof, or to small houselike structures in which it is placed. A dormer may be built entirely in the roof or its face may be the continuation upward of the wall. The construction of them is widely variant, but there are two general types, the *flat-roofed*, in which case the roof of the dormer is a continuation of the house roof at a lower pitch, and the *gable-roofed*, the roof of this type being a small gable. In the former, the intersection of the two differently pitched roofs is not flashed, one being simply an extension of the other. The valley between the roof of the gable-roofed dormer and the main roof is, however, much sharper than in the other types, consequently the valley must either be flashed, or the courses of shingles must be continued from the main roof on to the dormer without a break, the valley being in this case fitted with a fillet to give a less abrupt turn.

STAIRS

Stair building was at one time considered an art in itself; indeed, it required a very skilled artisan to accomplish the framing and fitting of the elaborate and wonderful curves and twists which in the past were included in the plans of even small buildings whose chief charm should have been the simplicity of their interior lines. At present the simple staircase is much in vogue, probably because housebuilders are more appreciative of them, and the change in style has worked to the advantage of the ordinary carpenter, who finds that he can build a simple staircase as firmly and as well as he can build any other part of the house. Mills producing the interior trim for residences commonly keep a stair builder employed, for the reason that he can be kept busy at this one job continually.

A glossary of some of the terms used in stair building will enable us to discuss the details of construction more readily.

BUILDING CONSTRUCTION

Staircase is the term applied to the whole set of stairs, or series of stairs, including landings.

Flight, that portion of the stairs between landings, between floors, or between a floor and a landing.

The *rise* of a stair is the height from the top of one step to the top of the next.

The *run* is the horizontal distance between the face of one riser to the face of the next.

A *riser* is the vertical board beneath the tread.

A *tread* is a horizontal board forming a step.

A *nosing* is that part of the tread projecting beyond the riser and includes the small molding below.

Carriages are the rough timbers supporting the treads and risers; they are sometimes called *strings*, or *stringers*.

A *newel* is the heavy post supporting the balusters where the stairs begin.

Angle posts are the posts supporting the balusters where their direction is changed.

Winders are steps which come in the angle of the stairs when turning a corner.

A *landing* is a section of floor between successive flights.

Open stairs are stairs built between walls.

Aside from simplicity, the requisites of good stairs are safety and comfort, with a proper consideration for the harmony with the other interior fittings. For safety, there should be a landing every ten or twelve steps, but often in small residences, where the stairs include sixteen or seventeen risers altogether, this idea is disregarded, and the stairs are made in one straight flight. The length of the landing should be at least equal to its width. No flights of less than three stairs should be permitted in any building, since they are dangerous. Winders should be avoided as much as possible, for the variation in the width of each individual riser from the inside of the turn to the outer end of step is so great as to make the stairs inconvenient, if not dangerous. The considerations for comfort require that the rise of the step be not more than $7\frac{1}{2}$ inches, and not less than $6\frac{1}{2}$ inches, while the width of the tread should be 9 or 10 inches. The width of the stairs should not be less than $3\frac{1}{2}$ feet, though in small houses where space is essential, and in kitchen stairs, this may be decreased to 3 feet.

In designing a house, the number of the risers only should be given, leaving their exact height to be determined by the carpenter; for this height will vary somewhat from the height figured in the plans. The approximate location of each individual step should be given, however, with great care to the arrangement, so as to provide sufficient room for the total number of steps required, and for sufficient headroom. The minimum distance from the underside of the floor opening should never be less than $6\frac{1}{2}$ feet, and 7 or even 8 feet is much better. A good method of determining the amount of headroom is to use the front edge of the trimmer of the stair well as a center, and with a radius of 6 feet, strike an arc, from which the front edge of the tread should be kept clear.

Construction

There are two general methods adopted in stair construction, the one depending entirely upon careful fitting to produce a tight joint at the ends of the treads, and the other having the stringer recessed for each pair of treads and risers. The first method is known as the American or Boston method, and the details of the construction are as follows: upon the inside of the stringers is nailed a strip called a "horse," which has triangular pieces cut out along one edge so as to form recesses for the steps, as shown in Figure 48. Upon these horses the risers are first firmly nailed, then the treads are put on, both being cut very carefully and fitted with exactness so no crack between step and stringer is perceptible. The second method results in what is called the housed or English stairs; here each stringer is cut out to a depth of ¼ inch, the recess being the exact shape of the stairs, with enough additional cut out to permit of the insertion of a wedge back of the riser and another one below the tread. The construction is shown in Figure 49. One of the basic principles of stair building being that their construction should be left until all plastering is done, it is evident that the housed stairs cannot be built without leaving any plastering to be done beneath the stairs until the stairs are finished. The wedges holding the treads and risers are inserted first, then the lath and plaster is applied, the dampness accompanying

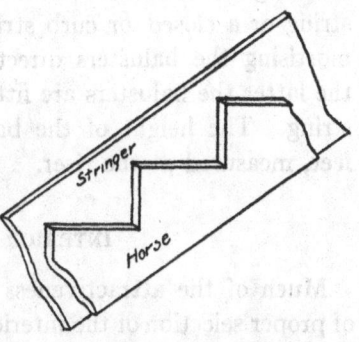

FIG. 48. — Boston stair construction. No cutting of stringer is necessary.

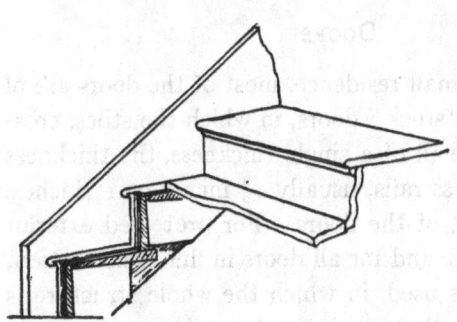

FIG. 49. — Housed stairs. Note position of wedges.

having a deleterious effect upon the finish lumber incorporated in the staircase.

On open stairs there is a wide variety of methods of finishing. The face of the stair may have either an open string or a closed or curb string, the former admitting of mortising the balusters directly into the tread, while in the latter the balusters are fitted into a shoe on top of the string. The height of the balusters should be about $2\frac{1}{2}$ feet, measured at the riser.

INTERIOR FINISH

Much of the attractiveness of a residence is the result of proper selection of the interior finish, and good workmanship in putting it up. The woods usually adopted for interior work are oak, birch, pine, red gum, chestnut, fir, and cypress; they are chosen principally because of their beauty, durability being secondary except in floors.

DOORS

In the ordinary small residence, most of the doors are of the type known as "stock" doors, in which the stiles, cross rails, and panels are all of a single thickness, the thickness of the stiles and cross rails, usually $1\frac{3}{8}$ inches or $1\frac{3}{4}$ inches, giving the thickness of the doors. For protected exterior doors in such houses, and for all doors in fine construction, the veneered type is used, in which the whole structure is composed of well-selected wood, glued together, and covered on the exterior with a thin layer of a more expensive wood. The latter types have the advantage of not being likely to warp, this being prevented by the construction; however, in an exposed place, especially exterior doors subjected to the action of the weather, veneered doors are not desir-

BUILDING CONSTRUCTION

able, inasmuch as dampness will ultimately cause the veneer to peel off, and the strips of softer wood underneath will swell and throw the door badly out of shape.

Interior doors are commonly 6 feet 8 inches in height, and 2 feet 8 or 10 inches in width, though closet doors may be of any size. Exterior doors are generally larger than interior doors, a common size being 3 feet by 7 feet. The thickness varies somewhat with the width, interior doors in ordinary house construction being $1\frac{3}{8}$ inches thick, while exterior doors are almost always $1\frac{3}{4}$ inches thick.

The vertical pieces of heavier lumber in a door are called the "stiles," the corresponding horizontal pieces, the "rails." These pieces frame the "panels" which may be from $\frac{3}{8}$ inch to $\frac{5}{8}$ inch thick, and their panel edges may be either ogee in shape, or as in the craftsman design, square.

The finish around a door opening, and window openings as well, is called *trim*, *casings*, or *architraves*, the second term being probably the more common. The various methods of arranging the casings are so numerous that they cannot be taken up here except to say that the simplest arrangement and design will probably prove the best, in appearance, neatness, and ease in caring for. When the casing is somewhat thin, it may not fit well with the baseboard; in this case a block, slightly wider and thicker than both the casing and baseboard, and about a foot long, is fitted in next to the floor, as a continuation of the casing. This block is designated as the *plinth*, or *plinth block*.

WINDOWS

Windows for residences are almost always of the vertical, sliding variety or else of the casement style. The former are usually made in two parts, the upper half sliding back of the lower part. At the middle of the window, or where the upper and lower sash meet, two constructions are used, one the *plain rail*, Figure 50, in cheap windows, and the other, or *check rail*, Figure 51, in the better class of work. *Casement* windows are made in one piece, and instead of sliding vertically in the frame, swing out, similar to an exterior door. Casement windows are much used in closets,

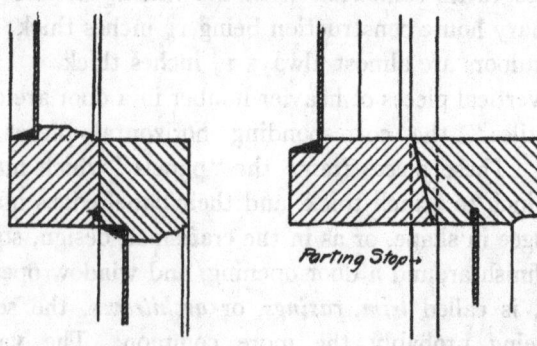

FIG. 50. — Plain rail window.   FIG. 51. — Check rail window.

high windows in dining rooms, etc.; their construction precludes the possibility of having exterior screens, so interior screens swung on hinges are employed, the wood used in the screens being identical with the interior finish. *French* windows are sometimes used where the windows open out upon a veranda; they are made like a pair of swinging doors, but on account of the necessarily weak framing, do not hold their shape well.

The finish of windows is very similar to that of doors, with the additional finish necessitated by the fact that windows do not reach to the floor. A flat horizontal piece extending at least $3\frac{1}{2}$ inches from the sash, and acting as an inside stop for the sash, is called the stool; and beneath it, flush with the casing at the side, is fitted a strip called the apron.

Other Finish

Around the bottom of the walls is fitted a strip of 8-inch stuff known as the baseboard, which should be very simple in design, since any extra beading, molding, etc., along the upper edge increases the difficulty of keeping it clean. In the corner formed by the floor and the baseboard is set a quarter-round mold, to prevent dirt getting into the corner. It is advisable to wait a year before setting this mold, since the joists may shrink somewhat and drop the floor without dropping the mold, leaving an unsightly crack. If this is remedied by lowering the mold subsequent to the application of stain or varnish on the base, a sort of line is exposed which is very difficult to eliminate by further treatment.

Sometimes the lower part of the wall is finished with a wainscoting varying in height from 3 to 5 feet. This may be either plain or paneled, the latter method being commonly used in dining rooms, in which case it is usually 5 feet high and capped with a corrugated plate rail, instead of a rounded nosing and cover which usually caps a low wainscoting. The panel strips should not be more than 3 inches in width; the panels themselves may be filled with wood or plastered, or covered with burlap or stamped leather for very rich effects. If a dining room is not paneled, it is well to put a chair rail around the walls, at a height which will prevent the backs of chairs knocking against the plaster and disfiguring it.

The old-fashioned method of supporting pictures, mirrors, etc., on nails driven into the walls, has been superseded by the addition to the interior finish of a picture mold, a small mold with a depressed edge next the wall. The picture mold is usually from a foot to a foot and a half below the ceiling, though it may be placed so far up as to practically constitute a corner mold, with just enough clearance above to admit the picture hook.

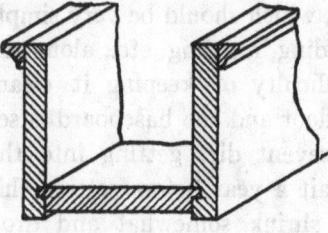

FIG. 52. — Beamed ceiling Better and cheaper than real beams.

Beamed ceilings are sometimes put in living rooms, dining rooms, etc., to secure certain effects, such as in Dutch or Mission interiors. The beams seen on the ceilings of dwellings are not usually solid, as they appear, but are a mere shell of thin stuff tongued and grooved together, one form of simple construction being shown in Figure 52. A half beam is usually placed around the

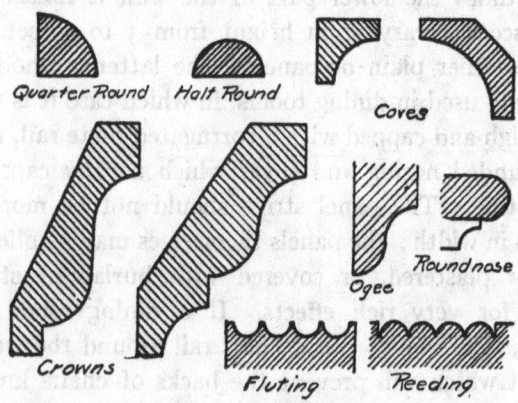

FIG. 53. — Moldings.

room, and the principal beams are fitted into this, the smaller beams being in turn framed into the larger ones.

Moldings are made in a great variety of sizes and shapes, and illustrations of some of these are shown in Figure 53.

Cupboards, bookcases, china closets, buffets, window seats, etc., are sometimes included in interior finish and can be installed at the time the house is built, of the same material and with the same finish as the rest of the interior woodwork. This can be done at comparatively moderate cost at the time, and adds much to the comfort and convenience of a home, to say nothing of the increased attractiveness, and the need for less movable furniture.

CHAPTER IV

ESTIMATING

To ascertain the probable cost of any extensive project, it is necessary that every detail of all the various divisions of the work be considered separately in regard to the cost of the production of the material, its conversion, and the labor required to prepare it and put it into the structure.

Other details, such as the material, its accessibility, cost of transportation, etc., are more local than general. For smaller structures, estimating is more easily done than for large ones, and with greater accuracy, since the total amount of construction of residences, etc., is so large that comparative costs can be justly estimated, even on the basis of unit construction.

The first requisite for a correct estimate is a complete set of plans and specifications. The specifications should be quite copious, giving in detail the grade and quality of all material used in the various parts of the structure, so that the estimator or contractor will be given no opportunity for "scamping" the work.

Let us suppose that the cost of an ordinary residence is to be ascertained. Beginning with the excavation, we shall take up the estimating in the same order as the materials will be put into the house. We should consider the site, so as to know the disposition of the earth that is to be removed, whether it is to be retained for subsequent grading, or hauled away. The time required for digging

ESTIMATING

and loading into a wagon or wheelbarrow 12 cubic yards of earth, is 9 hours. From this the cost can easily be figured, when the amount to be excavated and the excavator's wages are known.

Foundations are next considered. The total cubic content of the foundation wall is found, and if the wall is of stone, the number of cords, a cord being 128 cubic feet. A mason and his helper, using $2\frac{1}{2}$ bushels of lime and 5 bushels of sand, will lay about one cord of stone per day. Brickwork is figured by the cubic foot of 22 bricks, a workman and a laborer laying from 600 bricks in fireplaces and flues, to 800 bricks in walls, using about $1\frac{1}{2}$ bushels of lime and 5 times as much sand. As a rule, single-flue chimneys will cost about 40 cents per running foot, and double-flue chimneys about 70 cents.

The framing is next to be considered. Ascertain the linear measurement of the sills, and from their size and length calculate the number of feet, board measure. The same is done with the joists, for all floors, and with the studs and plates. A fair workman will fit in one day about 700 linear feet of sill or joint stuff, 800 linear feet of studding, and 500 linear feet of plate. In figuring studding, no deduction should be made for openings, as any overrun will be used in making diagonal braces, struts, etc. The number of rafters should be obtained, also their entire length, including cornice. Two carpenters will put in place about 700 linear feet of rafters on a plain roof, or 500 linear feet on one cut up by gables and dormers.

The amount of siding necessary is determined by finding the total exterior wall surface in square feet, including gables, and adding to this area $\frac{1}{4}$ for lapping. Two men will put up about 500 feet of siding in a day, this quantity being quadrupled in the case of plain vertical siding, such as barn siding.

When shingles are laid 4½ inches to the weather, 1000 shingles will cover approximately a square, or 100 square feet. A good workman will carry up to the roof and lay about 1500 shingles per day, and will use from 5 to 7 pounds of nails, according to the size used. If building paper is used under siding or shingles, this must be added, a roll of paper covering about 500 square feet, and requiring about one hour to lay and nail. Tin roofs are a little more expensive than shingle roofs, costing perhaps $1 more per square. Flashings in valleys, around chimneys, etc., are laid at the rate of about 30 linear feet per day, guttering at the rate of 60 to 75 feet. Cornices are built at about 30 feet per day.

In estimating the amount of material in floors, ascertain the number of square feet in the floor; then add ½ for waste and matching; this will vary somewhat with the width of the floor; measure openings, such as stair wells, hearths, etc., as if they were not there. The subfloor can be laid at the rate of about 800 square feet per day, the finish floor, requiring more care and time, being laid at the rate of 250 to 400 square feet per day, varying with the width of material, and whether or not it is tongued and grooved at both sides and ends. Plain baseboards, chair rail, etc., can be put in place at the rate of 150 feet per day, doubling this for sweeping and picture molds, etc.

Exterior doors require about 1 day to frame and hang, and a good workman will hang 3 or 4, even as many as 8, interior doors in one day, though 6 is an average man's capacity. One man will hang a sliding door in about 1½ days. Windows will cost about $3 or $4 apiece above the cost of the sash.

Plastering of good quality is done at the rate of about 25 yards a day, with a plasterer and laborer at work. In calculating the amount of plastering to be done, no deduction should be made for any openings at all. Exterior plastering, or stucco, will vary a great deal, but a good man and helper will put on 15 yards a day. Lath is sold by the thousand in bunches of 100, 1500 lath being required for 100 square yards of wall; a good lather will cover 50 yards of inside wall in one day.

A first-class staircase will approximate $4½ per step, including all material and labor. Cellar and back stairs will cost from $1 to $2 per step, according to width and finish.

The cost of painting and finishing is very hard to estimate, because there are so many governing factors. A gallon of priming coat will cover about 300 square feet, and subsequent coats will cover about 600 square feet. A good workman will paint 100 square yards in one day.

The cost of heating and plumbing is usually figured by the job; the cost of heating with hot air, steam, or hot water will in an ordinary residence run about $25, $40, and $45 per room, respectively. Plumbing will cost from $30 to $40 per room, depending upon the quality and elaborateness of fixtures. Electric wiring usually costs $1 per opening, every switch and light fixture being considered an opening.

The interior hardware, consisting of locks, latches, catches, handles, etc., must be figured piece by piece, though an allowance of $35 or $40 is sufficient to provide excellent design and workmanship, in a small house.

Cisterns, one of which is an essential part of every home, may be figured with reasonable accuracy by calculating the cost to be $1 per barrel capacity. Concrete can usually be figured at about 35 cents per cubic foot; this applies to sidewalks also, a 4-inch sidewalk with a 1-inch finish coat costing about 12 cents per square foot, a 3-inch one costing about 9 cents a square foot.

It is sometimes desirable and advantageous to be able to estimate the cost of structures quickly, or before the plans and specifications have been developed sufficiently to make an accurate detailed estimate. Two buildings built in the same locality should cost about the same per cubic foot, even though there be some difference in the size; it follows, therefore, that if we know the cost per cubic foot of different classes of buildings, in different localities, we can approximate quite closely the cost of any proposed building by multiplying its contents in cubic feet by the known cost per cubic foot of a similar building in the same locality.

The accompanying table for estimating small frame buildings is partially adapted from Kidder's Pocketbook:

FARM AND COUNTRY PROPERTY

Kind of Building	Cost per Cubic Foot
Dwelling, frame, no cornice	4c.
Dwelling, frame, small cornice	5 to 6c.
Dwelling, brick, cheap	8c.
Dwelling, good construction, frame	8c.
Dwelling, brick, good	10c.
Dwelling, hollow tile, good	9c.
Dwelling, modern frame	12c.
(In all of the above, basements are included)	
Barns, frame, plain	1½ to 2½c.
Barns, frame, well built	2½ to 3c.

Even a more rapid estimate can be made by the unit method, though it is a mere approximation and should never be held binding. The following table lists a few unit costs:

ESTIMATING

Per room in residences	$400	to $500
Per stall in horse barns	80	to 120
Per stall in cow stables	60	to 80
Per head in swine houses		
First-class construction	10	
Cheap construction	5	
Per head in sheep barns		
Well-built	8	
In cheap sheds	3	
Per fowl in poultry houses	.50	to 1.50
Per bushel in well-built cribs	.20	

CHAPTER V

DESIGN AND CONSTRUCTION OF FARM BUILDINGS

IN this chapter will be taken up a consideration of each of the various buildings found on the ordinary farms. The consideration will be made with regard to design and construction in general only, since no hard and fast rules can be laid down, each problem requiring a knowledge of local conditions for the most satisfactory and the most efficient solution.

GRANARIES

The chief fault of cribs and granaries as they are built in the majority of instances is that they are constructed with too little regard for strength and durability. A false economy is practiced when such a building is erected with just a few single stones or an occasional pier as a foundation, and with light, unsound timbers for sills and framing. One does not realize the tremendous strain to which the crib is subjected, especially at the floor and near the bottom of the walls. Assuming that the principles of hydrostatics hold, with certain restrictions and modifications, in the case of grain, an approximation of the amount of lateral pressure can be determined. This is equal to the area of a unit section of the wall (a section 1 foot in width and in height equal to the wall) multiplied by $\frac{1}{2}$ the height, and that product by the height of a cubic foot of the inclosed grain. Of course, friction would reduce the amount of lateral pressure, as would the relatively lower fluidity

CONSTRUCTION OF FARM BUILDINGS 119

of grain compared with water, to perhaps $\frac{2}{3}$ the theoretical pressure. On this basis the total lateral pressure on the wall of an oats bin 12 feet high and 16 feet long would be about 20,000 pounds, which is indeed considerable. It is evident that construction stronger and more secure than that in ordinary buildings erected simply for shelter is necessary in the case of granaries, which perhaps more than any other buildings on a farm are subjected to hard usage.

Every farmer has experienced the annoyance occasioned by wooden floors. Almost always inch boards are used; these will shrink in the summer when the bin may be empty, and when the new grain is poured into the bin, it promptly runs through the half-inch cracks; or the rats may have piled up such heaps of earth beneath the floor that it is continually damp, causing the boards to rot and the grain to mold. Occasionally a floor joist breaks, letting a section of the floor drop, with the result that a large quantity of grain is lost. All these unfortunate circumstances can be eliminated by the use of concrete floors, to which there still exist, in the mind of many farmers, serious objections, the principal one being that "the concrete draws dampness and makes the grain mold." All objections to concrete floors can be overcome by one precaution — make the floor right. Concrete, when properly made to suit existing conditions, is absolutely impervious to moisture, and can be kept as dry as any wood floor ever built. In addition to this, it is practically permanent, it makes an excellent base for the crib, and it always provides a smooth, sloping surface.

Concrete floors for cribs should be made an integral part of the foundation; that is, under each line of studding there should be a foundation wall perhaps 3 feet in depth, and 8 or 10 inches in thickness; between these walls, and as a continuation of them, there should be a 6-inch floor with a good subfoundation of well-tamped cinders or gravel. The floor should consist of a bottom layer of ordinary coarse concrete, with some accepted type of waterproofing mixed with it, and of a top layer composed of a rather rich mixture of cement and sand, also waterproofed. Such a floor as this is absolutely impervious, and if it is given a little slope to permit the drainage of any storm water that might be beaten into the crib, no better or more satisfactory floor can be devised.

To facilitate the emptying of grain from the cribs, sloping of floors is coming into great favor. A slope of $2\frac{1}{2}$ feet in 8 is sufficient to permit all the grain in a bin to slide out into drag belts or conveyors which carry the grain to the sheller or to wagons, as the case may be, without any hand labor being required, beyond that of keeping the conveyor from being blocked by too much grain. In the handling of ear corn this is an especially great advantage, effecting the saving of the labor of two or three men when the corn is shelled.

The labor of handling grain has become almost entirely mechanical. Formerly, the height of cribs was limited by the height to which a man could scoop grain, and on account of the labor and difficulty involved, cribs were seldom built more than 12 feet high. The advent of the modern small dump or elevator, sold at a price which made it a necessity to every farmer, and manufactured in such a multitude of variations for both inside and outside installations, makes it possible to have bins 20 or even 30 feet in height, and still permit the filling of the bins to be accomplished with a minimum of manual labor. The modern farmer builds cribs for permanence and convenience; he installs a vertical interior elevator which elevates his grain to the conveyors above, the conveyors distributing it to any corner of the crib desired; when the time comes for the removal of the grain, a few small doors are

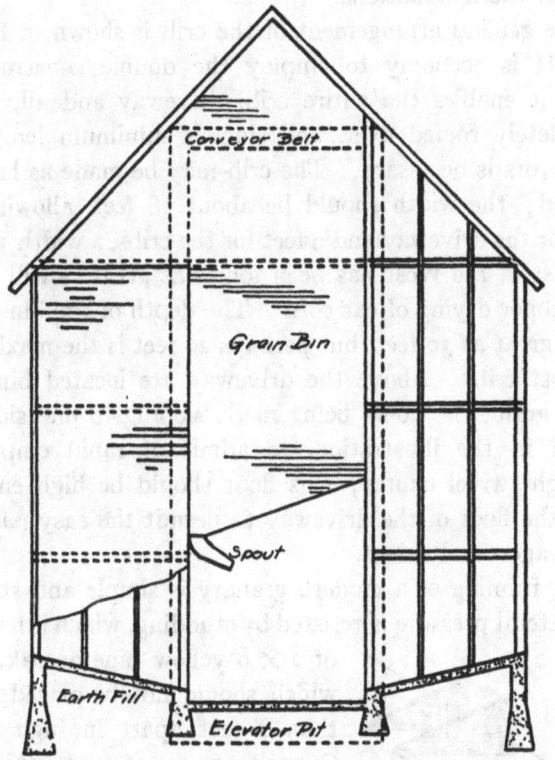

FIG. 54. — End view of granary, showing several modern construction features.

opened, the grain runs into other conveyors, and is carried to the sheller or to the waiting wagons, the whole procedure being carried on without the farmer raising his hand except to start machinery in motion. An added advantage of a complete elevator and conveyor installation is the facility with which damp grain may be transferred from one bin to another, thus aërating and drying it, and probably bettering its quality to such an extent that the higher price received for it when it is sold will soon pay for the cost of the installations.

The general arrangement of the crib is shown in Figure 54. It is economy to employ the double construction, since it enables the entire crib, driveway and all, to be completely roofed over, and since a minimum length of conveyors is necessary. The crib may be made as long as desired; the width should be about 28 feet, allowing 10 feet for the driveway and 9 feet for the cribs, a width which at least in the West has been found as great as will allow the proper drying of ear corn. The depth of the bins may be as great as 30 feet, but perhaps 20 feet is the maximum of most cribs. Above the driveway are located bins for small grain, the floors being made sloping to one side, as shown in the illustration, to admit of rapid emptying through swivel chutes; this floor should be high enough from the floor of the driveway to permit the easy passage of a wagon and driver.

The framing of a modern granary is simple and strong. The lateral pressure is resisted by studding, which is usually of 2 × 6 yellow pine or oak, and which should not be placed more than 2 feet apart in high bins. Opposite pairs of studs are tied with 2 × 6 ties securely spiked, two ties being necessary for a 20-foot stud. The studs are fastened at the bottom either to sill plates bolted to the concrete or by a special Goetz hanger, shown in Figure 55. There have recently been

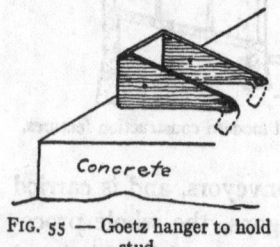

FIG. 55 — Goetz hanger to hold stud.

CONSTRUCTION OF FARM BUILDINGS 123

placed upon the market stud sockets made especially to be used in granary construction; these are made of cast iron, and are inserted into the concrete while it is still wet and pressed down level with the surface, making an admirable fastening. The latter form is simpler and does not interfere with the flow of grain. The two inner rows of studs are doubled below the floor of the upper bins, for the heavy weight of the grain above, coupled with the lateral pressure of the side bins, is likely to cause buckling of the studs unless they are quite strong.

The siding for corncribs is usually 1 × 6 rough fencing, put on with a 1-inch space between adjacent boards, so as to allow free passage of air. Small grain bins must be sided up closely, and for this purpose either ship lap or drop siding is used, the latter being preferable.

Some attempts have been made to construct an all-steel corncrib, using channel beams for sills and woven wire for siding, but with no appreciable degree of success. The difficulty in this construction lies in the fact that the woven wire rusts and deteriorates so rapidly that it cannot long withstand the lateral pressure.

MACHINE SHEDS

One great source of loss to the farmer is that resulting from lack of care of his farm machinery. Exposure to the weather results in the rapid deterioration of a machine, with an accompanying loss of efficiency, so that only a half or perhaps only a third of the value of the machine is finally realized. The average life of a grain binder, a complex machine requiring a rather heavy investment, is actually less than five years; experience has shown that its life may be easily prolonged to at least fifteen years with proper care. In the state of Illinois there was in

1910, $75,000,000 worth of farm machinery; on the basis noted above, $15,000,000 worth must be renewed every year. To properly house the machinery on the 200,000 farms of 50 acres or more, would require about $40,000,000, which would be saved in four years by extending the life of machines to fifteen years. Since with a little care a machine shed should at least last twenty years, the total saving in that term of years would be $160,000,000, a tidy sum to be distributed among the farmers of the state.

On the ordinary farm, the machine shed should be as simple as possible — a plain stucture with sufficiently wide doors to permit of the removal and return of implements with the minimum amount of time and labor. The interior should be clear of any vertical posts — this will require that the building be narrow or that a trussed roof be used, and should preferably have a concrete floor, though this will increase the cost of the structure by a considerable percentage, the concrete floor costing about one fourth of the whole building. If desired, a small shop may be included at one end of the shed, and will prove a wonderful convenience.

If a concrete floor is not used, rough 2-inch plank should form the floor, since the heavy machinery would soon destroy anything lighter. This should be supported on 4×6 sills, laid on a concrete foundation or on frequent brick piers. Horizontal siding will require that studs be used in the walls and 2×4 studs, doubled at the corners and at openings, and placed at 2-foot intervals, will be sufficient to make a firm wall and to support the roof. Should vertical siding be employed, a system of framing should be designed, using 4×6 posts and 2×4 or 2×6 nailing girts. It is advisable to cover the vertical cracks which appear between the boards with $2\frac{1}{2}$-inch battens.

CONSTRUCTION OF FARM BUILDINGS 125

The width of the building governs to a large extent the method of roof framing. Machine sheds adapt themselves readily to certain widths, 18 feet and 26 feet being perhaps the widths that can be most economically utilized. The floor plans following illustrate possible arrangements of machines within the buildings. If only an 18-foot width is used for the structure, the rafters themselves, with perhaps a collar beam or cross tie, are sufficient to support the roof. If a greater width is employed, a simple truss, like the one in Figure 56, must be built up, and placed at intervals of 9 or 10 feet. Ofttimes the collar beams and cross ties are used to support poles, lumber, and odds and ends that accumulate, and the weight of these things will give the roof a tendency to rack or sag. If the intention is to use the collar beams and ties for this purpose, the framing should be made extra strong to resist the additional strain.

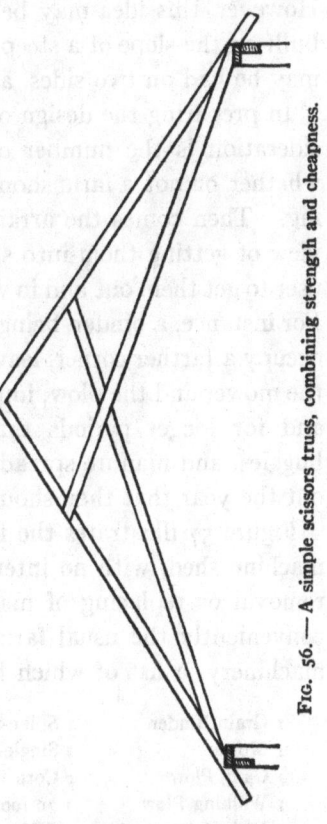

FIG. 56.—A simple scissors truss, combining strength and cheapness.

The floor of the shed should not be very high, and the approaches to the doors should be quite gradual, for otherwise it will be very difficult to run some of the heavier machines into the shed. Some builders advocate the use of two-story structures, but this is impracticable for the ordinary farm, and the added expense for the necessary hoist and the trouble of operating it would make it undesirable. However, this idea may be well worked out if the shed is built on the slope of a steep hill, so that natural approaches may be had on two sides, and a hoist will be unnecessary.

In preparing the design of a machine shed, the first consideration is the number of machines to be housed, and whether or not a farm shop is to be included in the building. Then comes the arranging of the machines with the view of getting them into such locations as to enable the user to get them out and in with the least amount of trouble. For instance, a binder, being used just once a year, may well occupy a farther corner, leaving the space near the door for the mower and the plow, implements which are used oftener and for longer periods than the binder. The wagons, buggies, and manure spreaders are used so much throughout the year that they should be especially accessible.

Figure 57 illustrates the floor arrangement of an 18-foot machine shed, with no interior posts to interfere with the removal or replacing of machines. It shelters easily and conveniently the usual farmer's equipment in the way of machinery, a list of which follows:

1 Grain Binder	1 Spike-tooth Harrow	1 Wagon
1 Mower	2 Single-row Cultivators	1 Spreader
1 Gang Plow	1 Corn Planter	1 Single Buggy
1 Walking Plow	1 10-foot Drill	1 Double Buggy
1 Disk Harrow	1 Self-dump Rake	

This shed has in addition an 8-foot shop across one end. The section devoted to buggies has a single sliding door of its own, because of the large amount of usage the buggies receive; this door arrangement permits the easy removal of the plow. The spreader and wagon may be located in

CONSTRUCTION OF FARM BUILDINGS

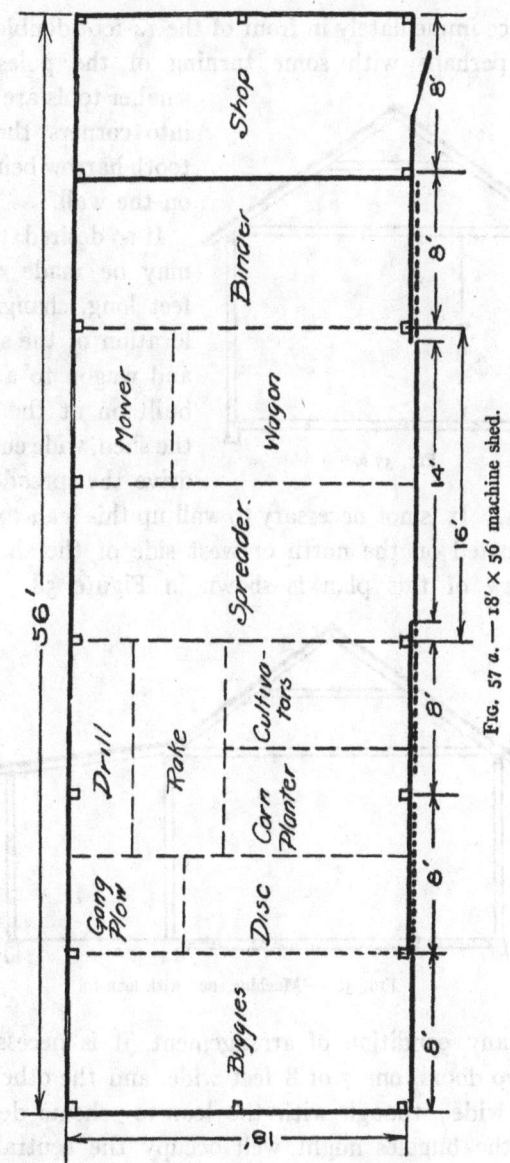

FIG. 57 a. — 18′ × 56′ machine shed.

the space immediately in front of the 14-foot double sliding doors, perhaps with some turning of the poles. The smaller tools are slipped into corners, the spike-tooth harrow being hung on the wall.

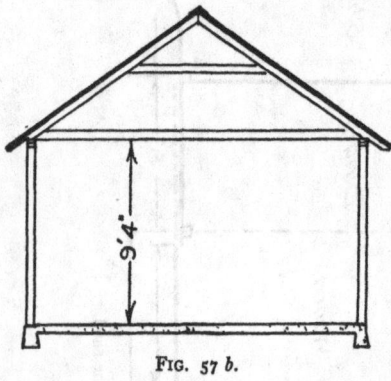

Fig. 57 b.

If so desired, the shed may be made only 40 feet long, changing the location of the spreader and wagon to a lean-to built on at the rear of the shed, wide enough to drive the spreader clear through. It is not necessary to wall up this lean-to, unless it is located on the north or west side of the shed. An end view of this plan is shown in Figure 58. Under

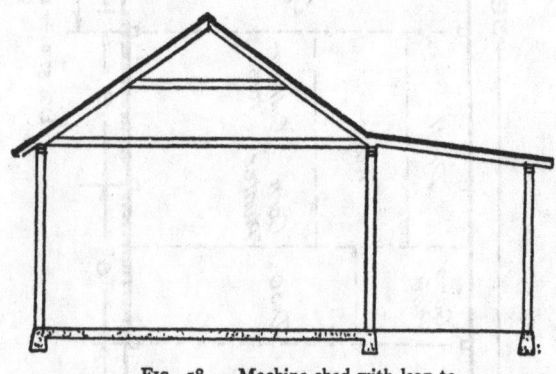

Fig. 58. — Machine shed with lean-to.

almost any condition of arrangement, it is necessary to have two doors, one 7 or 8 feet wide, and the other 12 or 14 feet wide; though with the lean-to scheme described above, the buggies might well occupy the central space

CONSTRUCTION OF FARM BUILDINGS

before the wide doors, thus obviating the necessity of other ones. Since the wide doorway must be kept clear, the eaves above must be supported by a stiff truss, which

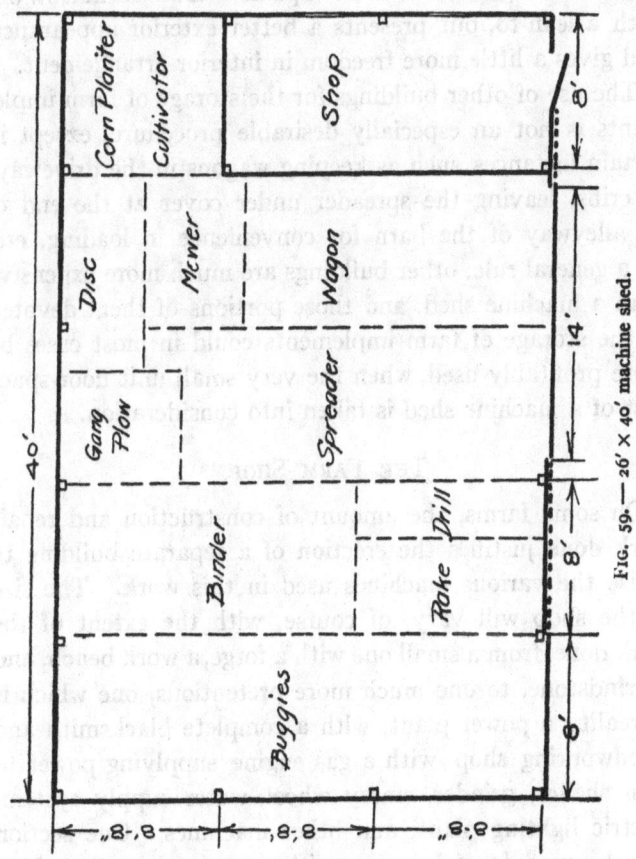

FIG. 59. — 26' × 40' machine shed.

will also carry the sliding door track; if the opening is not trussed, the roof will sag, and the doors will not run horizontally, but will drag and catch on the ground.

An arrangement for the same implements in a machine

shed 26 feet wide is shown in Figure 59. A shop, 8 × 16, is included in the arrangement; a larger one could be provided by simply extending the building to the length desired. The wider building is a little more expensive than the narrow one with a lean-to, but presents a better exterior appearance, and gives a little more freedom in interior arrangement.

The use of other buildings for the storage of farm implements is not an especially desirable procedure, except in certain instances such as keeping wagons in the driveways of cribs, leaving the spreader under cover at the end of an alleyway of the barn for convenience in loading, etc. As a general rule, other buildings are much more expensive than a machine shed, and those portions of them devoted to the storage of farm implements could in most cases be more profitably used, when the very small unit floor-space cost of a machine shed is taken into consideration.

The Farm Shop

On some farms, the amount of construction and repair work done justifies the erection of a separate building to house the various machines used in this work. The size of the shop will vary, of course, with the extent of the work done, from a small one with a forge, a work bench, and a grindstone, to one much more pretentious, one which is in reality a power plant, with a complete blacksmith and woodworking shop, with a gas engine supplying power to corn sheller, grinder, emery wheel, water supply system, electric lighting plant, and other machines. One section may be used for a laundry, with a power-driven washing machine, wringer, and mangle; another may shelter the cream separator and churn. If the shop is small, the construction of it should be quite simple, similar to the machine shed, with the exception that the walls should be made

FIG. 60. — A simple machine shed.

tight, for the shop will be used to some extent during cold weather, and the walls should be made close enough to retain heat. If the building is to be made larger, the framing must be rather heavy, since it will have to support overhead shafting and pulleys, with the attendant vibration and strain upon them. The building should have plenty of windows to give an abundance of light; the walls should be covered on the exterior with building paper and drop siding, and on the interior with Number 1 ship lap.

For the progressive farmer, a power plant is fast becoming a necessity. The equipment of the repair shop should include a good concrete forge with a hand blower, an anvil, a work bench with a heavy vise, a grindstone, an emery wheel, and a drill press. The forge should be placed along an outside wall, to get plenty of light, and should be set away from corners, to give plenty of room for manipulating long stock. A woodworking bench is a great convenience, and if installed should be located far enough away from the forge to obviate any danger from sparks and to keep out of range of soot.

The sheller and grinder may well be in a separate room, as they are always productive of more or less dust and dirt. Small bins, for the storage of grain, may be included in this part also.

In order that the engine may be kept in the best condition, it, too, should be located in a separate room, one that may be easily accessible from any part of the rest of the shop. The dust from the workshop, and from the feed room, would be injurious to the engine, and especially to a generator, should there be one. Since a rather large engine is used to drive some of the heavier machines, it may be economy to have a smaller engine to operate the laundry and creamery machines, and it may even be advisable to have them in a separate building more closely adjacent to the residence. Then the main power plant could be located as a part of the group of the other farm buildings, and be central to them.

Ice Houses

The great variety of uses to which ice is now put in the economies of living is sufficient reason for taking up a discussion of the principles of ice storage and preservation and the construction of ice houses. Before the manufacture of artificial ice became a commercial possibility the storage and distribution of natural ice were the only means of relief for residents of cities from the summer heat, and in the country, unless natural ice was at hand, the difficulty was equally great. An ample supply of ice is of greater economic importance in the country than in the city residence, for city people can purchase perishable supplies as needed, but the remoteness of country homes from markets often renders it necessary to use canned, corned, or smoked meat products during the season of the year when the table should be supplied with fresh meats. Not only is ice valuable and appreciated because of its use in preserving fresh meats, butter, and other table supplies, but the production of high-grade domestic dairy products is almost impossible without it. Many markets to which milk is now shipped demand that it be cooled before shipment to a degree not attainable without the use of ice.

The source of ice supply will vary with local conditions. Nature often supplies an abundance of ice from lakes, rivers, or large streams without any special plan on the part of man. Sometimes the water of a small stream or spring can be dammed up sufficiently to afford a water surface large enough to provide the desired amount of ice.

The stream or pond from which the ice is taken should be supplied from a source which is free from pollution or contamination, and from vegetation which, freezing in the ice, would be deposited as the ice melted in the refrigerator, rendering it unnecessarily filthy and dangerous to health. It is impossible to have pure ice unless the pond or stream is clean and the water free from contamination.

Principles of Ice Storage

The following principles, physical and mechanical, must be considered in the construction of a building in which to store ice: (1) to prevent ice from melting, it must have a minimum of surface exposed to the air or packing material; this is best accomplished by cutting the ice in as large cubes as can be conveniently handled; (2) the ice must be thoroughly insulated, to protect it from external influences, such as heat and air; (3) drainage is important because the lack of it interferes with insulation; (4) packing of the ice must be done carefully so as to prevent as far as possible the circulation of air around it.

Types of Ice Houses

Any such advantages as are offered by shade and exposure should be taken advantage of in the location of an ice house, since at best ice is a highly perishable product requiring special equipment for its preservation. A shady location, with a northern exposure, is decidedly better than any other.

With reference to general design, ice houses are of three types: (1) those built entirely above ground; (2) those partly above and partly below; and (3) those entirely below ground. As a rule, the first type can be more easily and economically built than any other, because no excavations, which are expensive to make and difficult to insulate and drain properly, are necessary. It might at first be considered advisable to take advantage of the comparatively even temperature of the earth and its apparent coolness, but when we realize that the temperature of the earth at a depth of 5 or 6 feet is about 55 degrees F. the year around, it becomes evident that the stored ice must be protected from earth heat as well as air heat. It is easier, of course, to fill a partly subterranean house with ice, but this advantage is more than counterbalanced by the difficulty in removing it as it is used.

The Construction of Ice Houses

Two important considerations in the construction of any ice house are the character of the insulation and the cost of construction. The climatic conditions must also be considered, and the probable amount of ice that will be necessary. A ton of ice occupies approximately 35 cubic feet. Four or five tons are usually all that a single family will use during a season, so if the ice is to be for private use only, it is desirable that several families unite in putting up their ice supply together, if this can be accomplished without inconvenience.

An inexpensive ice house which will serve quite satisfactorily in the region whose climate is similar to that of Chicago or New York can be constructed as follows: choose a site that is thoroughly well drained; if the area is not drained naturally, grade the surface so that no water can ever flow into or through the building, and so that water from the melting ice can be quickly disposed of. Having provided for the disposal of the water, both from within and from without, set a series of 2 × 4 posts around the four sides of a square of the dimensions desired; a building 10 feet square will allow of a storage capacity 7 feet square. Board up the inside with ship lap, and the outside with ship lap or drop-siding. The space between

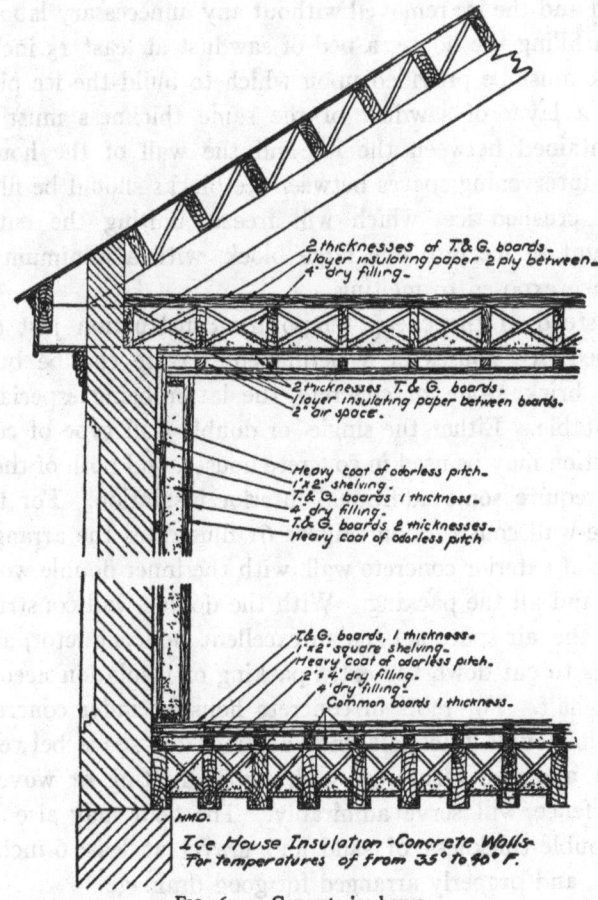

FIG. 61. — Concrete ice house.

the inner and outer walls should be filled with some perfectly dry material, like sawdust or packed shavings. The roof may be a simple gable or hip roof, with common shingles nailed on, and with a little ventilator cupola provided in the peak. A continuous door, similar to a silo door, should be built in one side, in order that the house may be filled and the ice removed without any unnecessary labor.

In filling the house, a bed of sawdust at least 15 inches thick must be provided upon which to build the ice pier; and a layer of sawdust of the same thickness must be maintained between the ice and the wall of the house. Any intervening spaces between ice blocks should be filled with crushed ice, which will freeze, uniting the entire amount of ice into one large block, with a minimum of surface exposed to melting.

Instead of the cheap temporary construction just described, ice houses of a permanent nature can be built from brick, stone, or concrete, the latter being especially adaptable. Either the single- or double-wall type of construction may be used in concrete houses, and both of them will require some additional interior insulation. For the single-wall construction, Figure 61 illustrates the arrangement of exterior concrete wall, with the inner double wood one, and all the packing. With the double-wall construction, the air space acts as an excellent nonconductor, and serves to cut down the extra packing or insulation needed by a half. For roofs in concrete houses, cinder concrete laid in double thickness with a 2-inch air space between them, and well reënforced with strong netting or woven-wire fence, will serve admirably. The floor may also be of double thickness, of sand and gravel at least 6 inches thick, and properly arranged for good drainage.

The Silo

A silo is an air-tight, water-tight tank, in which green, succulent herbage may be placed and preserved, very much as fruits are preserved in glass jars. Just as the housewife finds that it is in those jars which were not air-tight that the fruit does not keep well, so does the farmer find that the admission of any air to well-packed silage will result in mold and decay. The silo may be and often is wholly above ground, but very frequently has a large part of its total capacity below ground; its original form was that of a large pit, being entirely below ground. From this there developed the square silo built above ground, but this was a failure, for several reasons; it could not be built economically, nor strong enough to resist the internal pressure, nor did the silage keep well in it. The settling of the silage was uneven, due to the greater amount of friction in the corners of the walls in proportion to the weight of silage than in the middle of the structure. Following the square silo came the octagonal one, but this possessed many of the inherent defects of its predecessor. Finally, there was evolved the round silo, which has been universally accepted as the one shaped most perfectly, both from the theoretical and the practical viewpoint.

Some of the essentials of a good silo are as follows:

(1) A location which is at once sheltered from cold winds and easily accessible, both for filling and for emptying. As to whether an inside or outside location is the better, depends more than anything else on the type of the barn. A circular barn can very well have the silo located in the center, where it will act as a splendid support for the framing of the barn itself. Under almost any other circumstance, it is better to have the silo outside of the barn, on account of the strong, pungent odor of the silage, which is often disagreeable, and because of the fact that it occupies a great deal of floor space which might otherwise be more profitably used.

(2) A foundation which is thoroughly solid and substantial, extending at least 4 or 5 feet below the ground, so that the bottom of the silo is below the frost line. It is quite necessary that this be well drained, in order to obviate as far as possible any settling, which is likely to occur with the heavy structure above.

(3) Walls which are smooth, strong, straight, and solid. The lateral pressure on silo walls is something enormous, and the walls must be sufficiently rigid to resist this pressure as well as to resist storms. Since air causes silage to spoil, there must be no pockets or depressions in the wall in which air can be imprisoned, and which will prevent the even settling of the silage.

(4) Some protection for the top of the silo, so that silage may not decay at the top. Some authorities claim that a roof is an unnecessary expense, and that silage will spoil to just as great a depth at the top with a roof as without one. This claim is hardly substantiated, but some owners of silos simply sow some oats or rye on top of the silage, and claim that the thick sod resulting protects the silage adequately.

Size of the Silo

The capacity of a silo is determined by the number of cattle to be fed and by the length of the feeding period for silage. This period usually lasts about 200 days, though some stock raisers find silage a profitable and satisfactory feed for the entire year.

Silage, to be used advantageously, must be fed off the top of the silo; any opening in the bottom would admit air and cause the silage to spoil. Even when feeding from the top, unless it is begun just as soon as the silo is filled, there will be some loss, as a thin layer at the top exposed to the air will decay and must be discarded. The silage, after feeding has begun, must be removed at the rate of at least 2 inches per day to prevent the formation of mold, and the diameter of the silo should be so gauged as to insure that this amount will be consumed. Unless the silo is constructed so as to be particularly impervious to cold, there will occur some freezing of the silage around the walls; the frozen silage, if fed just as soon as it is thawed, will not have any particularly different effect upon cattle than silage which has not frozen, beyond a mere laxative action. In some sections of the country, it is the practice to make the surface of the silage cone-shaped, this preventing to some extent the freezing that sometimes occurs around the edges near the top.

The accompanying table gives the capacities of silos required to supply silage to herds of different sizes, fed either for 180 or 240 days; the corresponding correct diameter is also included. Though diameters of 22 feet are given, 20 feet should be the maximum, since any greater diameter means an excess of labor in removing the silage.

Number of Dairy Cows	Feed for 180 Days	Feed for 240 Days	Diameter of Silo
8	29 tons	40 tons	8 ft.
10	36 tons	48 tons	10 ft.
15	54 tons	72 tons	10 ft.
20	72 tons	96 tons	12 ft.
25	90 tons	120 tons	14 ft.
30	108 tons	144 tons	16 ft.
35	126 tons	168 tons	16 ft.
40	144 tons	192 tons	18 ft.
45	162 tons	216 tons	18 ft.
50	180 tons	240 tons	20 ft.
60	216 tons	288 tons	22 ft.
70	252 tons	336 tons	22 ft.
80	288 tons	384 tons	22 ft.
90	324 tons	432 tons	22 ft.
100	360 tons	480 tons	22 ft.

The height of silos should be as great as can be obtained with the most economical construction, bearing in mind, however, that for silos from 8 to 10 feet in diameter, the height be not more than 40 feet, that for silos from 12 to 15 feet in diameter, the height be not more than 45 feet, and that the height of any larger silo be not more than 60 feet. Silos of a greater height than this are more or less inaccessible, and the filling of them is accomplished with more or less difficulty. For this reason two silos of less diameter and height are often preferable to one large one.

In estimating the capacities of silos, consideration should be given the fact that the silage at the bottom of a silo is so much more compactly compressed than at the top that its unit weight is much greater than that of the silage at the top. For instance, at the bottom of a silo 36 feet deep, the weight of a cubic foot of silage is approximately 60 pounds, while the same amount of top silage weighs but 18 or 20 pounds. From this it is seen that the capacity increases in greater proportion than does the depth.

In a bulletin issued by the Wisconsin experiment station, a table reproduced on page 141 indicates the weight of silage at different distances below the surface, and the mean weight of the silage for silos of different depths.

The lateral pressure tending to burst a silo is considerable, and all silos require some sort of reënforcement to resist this strain. Professor King of Wisconsin has found that the pressure on the walls of the silo due to the weight of the silage is 11 pounds per square foot of wall area per foot of length. That is, at a depth of 10 feet, the lateral pressure per square foot is 110 pounds; at a depth of 20 feet, it is 220 pounds; at 30 feet, it is 330 pounds.

Depth of Silage	Weight of Silage at Different Depths	Mean Weight of Silage per Cubic Foot	Depth of Silage	Weight of Silage at Different Depths	Mean Weight of Silage per Cubic Foot
Feet	Lbs	Lbs	Feet	Lbs.	Lbs.
1	18.7	18.7	19	45.0	32.6
2	20.4	19.6	20	46.2	33.3
3	22.1	20.6	21	47.4	33.9
4	23.7	21.2	22	48.5	34.6
5	25.4	22.1	23	49.6	35.3
6	27.0	22.9	24	50.6	35.9
7	28.5	23.8	25	51.7	36.5
8	30.1	24.5	26	52.7	37.2
9	31.6	25.3	27	53.6	37.8
10	33.1	26.1	28	54.6	38.4
11	34.5	26.8	29	55.5	39.0
12	35.9	27.6	30	56.4	39.0
13	37.3	28.3	31	57.2	40.1
14	38.7	29.1	32	58.0	40.7
15	40.0	29.8	33	58.8	41.2
16	41.3	30.5	34	59.6	41.8
17	42.6	31.2	35	60.3	42.3
18	43.8	31.9	93	61.0	42.8

Types of Silos

Almost every sort of material that can be used in construction has been incorporated at various times into silos. Wood, brick, hollow tile, stone, steel, concrete, all have been used with more or less success, but a gradual elimination has resulted in leaving three materials from which in the present time the great majority of silos are built; these are wood, hollow tile, and concrete; steel is used for reënforcement and support in all of the silos built from these materials.

Variation in the method of using the building materials gives rise to a broader classification, and it is this classification that we shall follow in discussing the construction advantages, and disadvantages, of different kinds of silos. Thus we have the stave silos, all practically the same, differing only in minor details; the Gurler silo, originally of wood and plaster, now also manufactured entirely of steel and plaster; the hollow tile silo, with minor variations as to shape of block, etc.; and the concrete silo, which may be made of solid or hollow blocks, or of single- or double-wall monolithic construction.

The Stave Silo

The stave silo is made of long staves usually 2×6 inches, held together in a barrel-like form by hoops of steel. The popularity of this form of silo is evidenced by the thousands of them in successful use throughout the United States. Its main advantages are its comparative cheapness, especially where only a temporary silo is desired, and its portability, the construction being such that it may be dismantled and removed to another location with the loss only of the foundation.

There was a time when common 2×4's or 2×6's were used as staves, setting them up on end and drawing them together with iron bands or hoops. There was so much leakage through the cracks between the staves that in a great many cases silage spoiled, and as a result the use of the silo was condemned. In modern practice, however, this possibility of leakage is to a great degree eliminated, because the staves are matched so carefully, both side and end, that the outlet of silage juices and the admission of air is practically prevented. In the cheaper forms of stave silos, the matching is accomplished by merely beveling the edges, and trusting to the closeness with which the staves are drawn together to keep the structure comparatively water-and air-tight.

The construction of a stave silo calls first for a good substantial foundation. If the silo is to be entirely above ground, a circular foundation of concrete or brick masonry, varying from 10 to 18 inches in thickness according to the size of the silo, and the sort of ground upon which the silo is located, must be built by digging a trench of the required width and deep enough to reach firm ground below the frost line; if the foundation is to be of concrete, circular forms should be erected above the trench so as to bring the concrete at least 1 foot above the ground. With a brick foundation, forms are of course unnecessary. Usually it is an advantage to have part of the silo below ground, and when this is desired, a pit of the required diameter and depth is dug, lining the sides of the pit with brick masonry or concrete. In any kind of silo construction, the floor should not be built integral with the walls or the foundation; for should such be the case, the weight of the silage would probably break up the floor badly, and possibly result in the failure of the silo. Some sort of a drain, extending outside of the walls, should be installed, to take care of any seepage beneath the silo.

With the foundation and floor in place, the erection of the staves is next in order. The wood from which staves are made may be any one of the following, their comparative value being in the order in which they are given: cypress, California redwood, white pine, cedar, fir, yellow pine. It is important that the wood be straight and free from any loose knots, sapwood, or any other serious defect; the material should be as uniform as possible, the staves being of the same width and thickness. The silo may be made of one-piece staves, but this is rather expensive, especially as the height increases, and just as satisfactory a silo may be made with built-up staves. When the stave is made of two pieces of different lengths, the joint should be made accurately, and splined together by making, in the ends to be joined, a saw cut 1 inch deep and parallel to the sides of the stave, and inserting a sheet-iron spline, preferably galvanized, as shown in Figure 62. The ends of the stave should be painted with white lead to further protect the joint. To fasten the adjacent staves together, half-inch holes are bored in the edges about 5 feet apart; these holes are made on one side only of each stave and must be staggered on adjoining staves; they should be about an inch deep in 4-inch staves, and 3 inches deep in 6-inch staves. The purpose of these holes is to allow spiking the staves together when set up, the spike being driven into the bottom of the hole, and passing through the rest of that stave and into the next one, as shown in Figure 63.

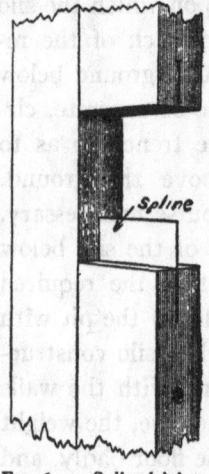

FIG. 62.—Splined joint.

Before the staves are put up, the number of doors must be decided upon, and their distance apart. The doors are usually from 2 to 2½ feet high, 2 feet wide, and may be from 2½ to 3 feet apart. The location of the doors is laid

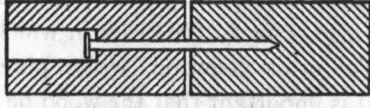

FIG. 63. — Method of fastening staves.

out upon a stave, and saw cuts are made halfway through it at an angle of 45 degrees, as shown in Figure 64. This stave is put up at one side of the place where the doors are to be cut, and after the staves are all up a saw is inserted in the saw cut already begun and a cut of the desired width of door is made. The purpose of the slanting cut is to make a door that can be removed only from the inside, so that when the silo is filled, the lateral pressure will hold it in place. The portions of the staves forming the doors can be cleated together after being sawed out, the cleats being placed on the outside of the door so as not to interfere with the settling of the silage.

Sometimes a continuous door is desired; this adds to the convenience of the silo, and is somewhat more difficult to construct. In putting in a continuous door in a stave silo a door frame should be provided of 4 × 6 pieces, held 20 inches apart by pipes, and kept from spreading by ⅝-inch bolts placed at intervals of 2 feet, running through the pipe, as shown in Figure 65. Iron washers should be placed between the ends of the pipe and the door frame, and under the heads and nuts of the bolts. The inside corner of each timber should be chamfered as shown in the illustration, to provide a shoulder against which the doors can rest. The doors themselves are made of double thickness of tongued and grooved flooring, with a thickness of tarred paper between, the inside layer of flooring running vertically and the outside horizontally.

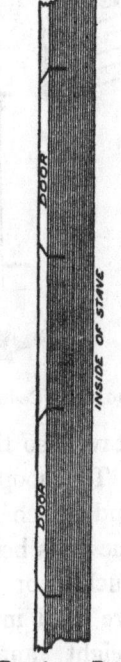

Fig 64 — Door location for a stave silo.

The first stave is placed with its inner face on the line, 5 inches from the inner edge of the foundation. It must be plumbed in both directions and securely fastened at the top and bottom by braces nailed to stakes firmly driven into the ground, or to some adjacent building. A movable stepladder may be used instead of scaffolding, and this may be moved along and kept in the right place from which to work. The next stave is then set up and nailed to the

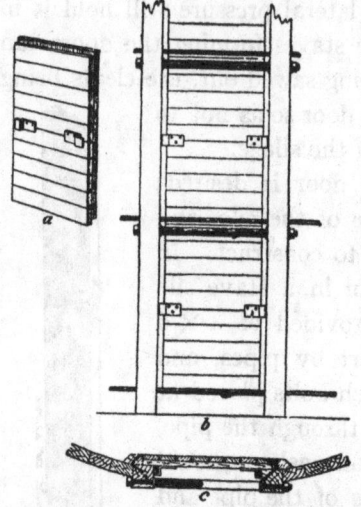

FIG. 65.— Continuous door for a stave silo.

first with 30d. or 40d. spikes, which are started in the holes previously bored (Figure 63) and driven home with a drift punch. Other staves are then erected until the place is reached where the doors are to be; the door stave, previously prepared, is then nailed in position and the remaining staves erected. In setting up splined staves, the long and short pieces should alternate in adjacent staves, so the joints may be staggered.

The hoops are next in order. These are made of $\frac{1}{2}$-, $\frac{5}{8}$-, and $\frac{3}{4}$-inch rods, in sections from 10 to 14 feet in length, the ends being threaded to admit of being joined to turn-buckles or lugs. The $\frac{1}{2}$-inch rods are used in the upper third of the height, and $\frac{5}{8}$-inch rods for the remainder. Should the silo be above 30 feet in height, the lower third should be of $\frac{3}{4}$-inch, the middle third of $\frac{5}{8}$-inch, and the upper third of $\frac{1}{2}$-inch rods. Two hoops should be placed below the first door, and two between adjacent doors all the way up, putting two hoops above the top door if this space is more than 2 feet; if less than 2 feet, one

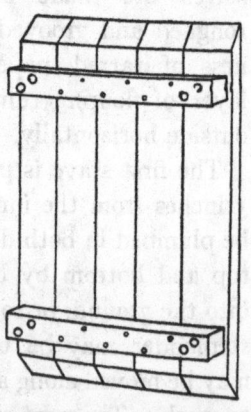

FIG. 66.— Completed door.

CONSTRUCTION OF FARM BUILDINGS 147

hoop will be sufficient. When the hoops are in place, they should be tightened up and staples driven 2 or 3 feet apart over each hoop to hold it in place, should it become loose.

After all the hoops have been placed in position, the doors should be cut out, using the stave previously cut as a guide. By cutting through 4 or 5 staves, a sufficient width is obtained for the doors. A completed door is shown in Figure 66.

Figure 67 shows the silo with all the staves erected and the hoops in position. One door has been cut out completely, and another one partially so, showing the method of doing this part of the work. The figure also shows some other details of construction, such as the staggering of the joints in the staves, the method of anchoring the silo to the foundation, the location of the guide stave for the doors, and the lugs on the hoops.

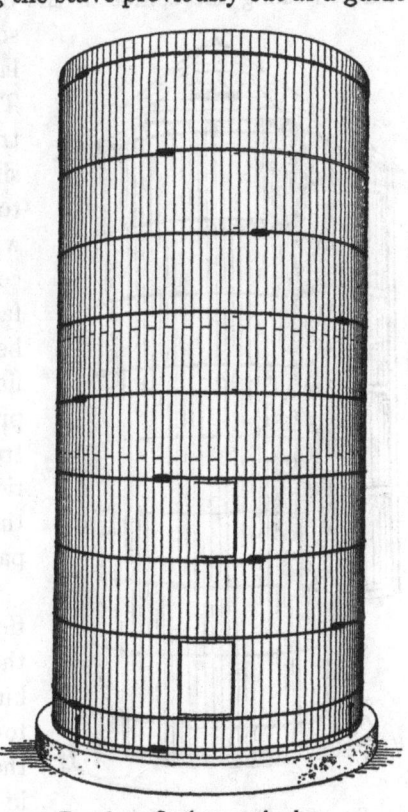

FIG. 67. — Sawing out the doors.

If a roof is put on the silo, it may be easily constructed with eight or ten sides, using 2 × 6 stuff for rafters, and employing shingles or prepared roofing for a covering.

For filling the silo, a door must be put in at the top; this door may be either a trap door in the slope of the roof, or a vertical door inserted in the face of a gable. Since the silo must be emptied from above, a ladder must be constructed along the course of the doors, and a chute built so as to inclose both the ladder and the doors. This is shown in the illustration of the finished silo, Figure 68. In order to properly preserve the wood of which the silo is constructed, the interior face of the staves should be given a coat of carbolineum or of some other preservative. The attractiveness of the exterior will be enhanced if the silo is kept well painted.

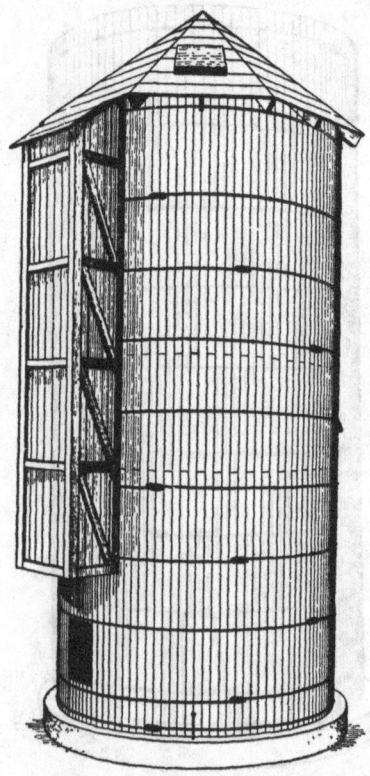

Fig. 68. — The completed silo.

The foregoing description is that of just one of the dozens of different kinds of stave silos built to-day. In all those on the market, the variation in construction is almost always in some detail, as a special door, or hoop, or anchoring, or reënforcement, or material used, the method of erection, etc., this detail being controlled generally by a patent. Stave silos will always be built to some extent, but the pronounced advantages of the better and more permanent types of silos are gradually limiting their use.

The Gurler Silo

The Gurler silo is the invention of Mr. H. B. Gurler, who built his first silo in 1897. Long and successful use of this type of silo has shown the practicability of this form of construction, the details of which follow.

The preparation of the foundation for the Gurler silo is exactly similar to that of an ordinary stave silo. Upon this foundation are set studs of 2×4 stuff, 16 inches on center, held in place vertically by an occasional girt, the studs resting on a 2×4 sill. Since it is usually difficult to obtain studs equal in length to the height of the silo, each stud may be made of two pieces, lapping them 2 feet. On the inner face of the studs is nailed the lining, which consists of $\frac{1}{2} \times 6$ stock made by splitting common fencing with a saw. On top of the studs is nailed a 2×4 plate.

Provision for doors is made by doubling the studding on each side of the row of doors and binding these together either by $\frac{5}{8}$-inch rods between the doors or by doubling the lining on the inside across the space between the doors and extending it six or eight feet on each side.

The silo is then lathed on the inside with a special form of lath, known as Gurler lath, which consists of the ordinary lath with the edges beveled. The lath is applied with its narrower face nailed directly to the lining, and spaced so that a dovetail joint is formed for the mortar. Sometimes the Byrkit lath, shown in Figure 41, is used to fill the place of both lining and lath.

The silo is then plastered on the inside with a cement plaster made in the proportions of 1 part of cement to 2 parts of sand. This should be applied in two coats, the first coat being roughened and left to dry thoroughly before the second coat is applied.

The exterior of the silo may be covered in any one of several ways. The covering may be of ordinary weatherboarding, bent around the silo and securely nailed, especially at the ends; however, this is not entirely satisfactory, for there is a continual tendency in the siding to straighten, and at the ends this tendency is strong enough to cause the joint to open when the nails become somewhat rusty and the wood a little rotten. The most satisfactory method of covering the exterior of the silo is to set horizontal girts into the studding at vertical intervals of about 3 feet, and to these nail vertical boards 6 or 8 inches wide, either plain boards with the cracks covered with battens or ship lap.

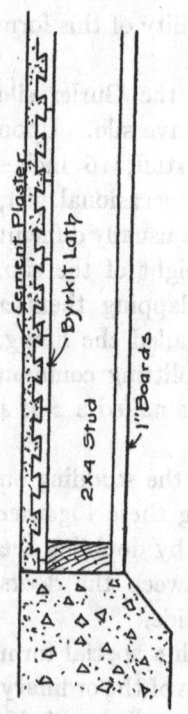

FIG 69. — Gurler silo construction, of the wood and plaster type.

The doors for the Gurler silo may be either individual or continuous, similar in construction to those of the stave silo. If a continuous door is used, the method herein described of binding the sides of the door frame together by means of double thickness of lining cannot be used; only the iron rods are practicable. The roof construction is similar to that of the stave silo.

In Figure 69 is shown a cross section of the walls of the Gurler silo, with all the details illustrated.

There has been evolved a special form of the Gurler

silo which apparently possesses all the good qualities of the original form in addition to that of permanence. In this form the framework consists of steel channels in place of studs, bound together internally and externally by hoops of strap iron, to which is fastened metal lath. The silo is plastered inside and out with cement plaster, the exterior being given any of the numerous stucco finishes. The roof is of concrete, making the whole silo a fireproof, permanent structure.

The Vitrified Tile Silo

The vitrified tile silo is a type of silo construction that will endure. This silo is built of blocks made of potter's clay, burned to a vitreous body, presenting a hard, smooth surface, impervious to moisture, and resistant to the action of the acid in the silage juices on the interior, and of the weather on the exterior.

The Agricultural Experiment Station of Iowa has done much valuable work along the line of investigating the value of the vitrified tile silo, and has evolved a type in which the hollow air space is retained, therein differing from the Grout silo, which has the hollow walls filled with concrete. It is claimed that the hollow wall of the Iowa silo tends to make it frost-resistant.

When properly constructed, the vitrified tile silo is undoubtedly an excellent one; the main advantages of it are the comparative ease with which it is built, the low maintenance cost, and its permanence.

In the construction of this type of silo, the foundation is first prepared so as to bring the bottom of the silo about 4 feet below the ground; this is for the sake of economy, as it is easier to work when near the ground than when 30, 40, or 50 feet above the ground, and less scaffolding is necessary. The foundation may be of a solid concrete wall 10 inches thick widening to 16 inches at the bottom; or it may be built of brick with proper footing; or it may be built of the vitrified tile itself, the bottom course being two 8-inch blocks laid flatwise side by side, then a course of blocks laid flat crosswise, completing the footing, the next course being the first wall course Sometimes a combination concrete and block foundation is used, in which concrete is used for the footing. The cement mortar in which the foundation blocks are laid should contain waterproofing, and should be used abundantly and carefully, to prevent the ingress of surface water, which might result, in the case of a hollow wall, very seriously should freezing occur.

The floor of the silo should be constructed similar to the way in which concrete sidewalks are built, or it may be made of the blocks themselves covered with a thin coat of plaster. While a floor is not absolutely necessary, it is desirable, since the portion of the silo below ground can be made more nearly water-tight, the floor can be kept clean, and no earth or dust becomes mixed with the silage.

The foundation having been laid, the next consideration is the wall. The blocks from which the walls of the Grout silo are laid up are shaped as shown in Figure 70, with curved faces, and part of the partitions cut out to form a groove in which to lay the reënforcement. Blocks with either a straight or curved face have been used in the construction of the Iowa silo, but the former have not proved satisfactory, since the straight-faced block when laid up in a circular wall makes a very irregular surface, especially

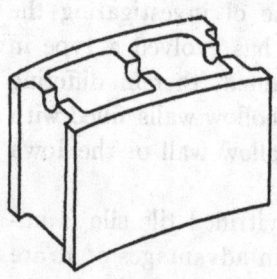

FIG. 70. — Vitrified silo block, used in Grout silo.

when a rather long block is used. It has been found that a block 12 inches long, 8 inches wide, and 4 to 6 inches thick is the most desirable size, since they are not difficult to handle and are laid up rapidly and require a minimum of mortar. In the Iowa silo no special provision is made for placing the reënforcing, as the largest reënforcement used is a No. 3 wire, which in diameter is less than the thickness of the mortar joint; therefore, it does not interfere with the laying of the blocks, and the mortar protects it from rust.

The mortar in which the blocks are laid up is composed of cement, lime, and sand. The lime is a necessary ingredient, as a mortar made of cement and sand alone is not plastic enough to adhere to the blocks. The amount of lime to be used should not be more than is necessary to make the mortar easily workable; a mixture of 1 part of cement, a little less than 1 part of lime, and 4 parts of good sand, medium fine, will make a mortar that any workman should be able to handle well.

It is in the method of laying the tile in the walls of the silo that the main difference lies between the Grout and the Iowa silos. In the former, the tile are placed on edge, with the hollow spaces extending vertically through the tile and through the silo; this permits of the placing of vertical reënforcing within the tile. When the silo is completed, the wall is filled with a slush concrete, which results in making the structure a solid, unified mass, absolutely impervious to air or moisture. The tile in the Iowa silo are laid on edge or flatwise, with the hollow space extending horizontally, no vertical reënforcing being used except near the opening for a continuous door.

The reënforcement of both the Grout and the Iowa silos consists of wire of various sizes according to the amount of bursting pressure to which the wall is likely to be subjected at various heights. The vertical spacing is, of course, controlled by the height of the blocks, since no space can be less than the height of a single block; near the top of the silo, where the lateral pressure is very small, reënforcement may not be inserted at every course.

In the accompanying table is given the amount of horizontal reënforcement necessary in a silo of this type, assuming the reënforcement to be placed between each course of 8-inch blocks. For each diameter from 12 to 20 feet, a double column is given, the figures in the first indicating the number of No. 9 wires to be used, and in the second, the number of No. 3 wires. Thus, at a depth of 20 feet, in a 16-foot silo, the reënforcement should consist either of three No. 9 wires or of one No. 3 wire. It is well to have the topmost reënforcement doubled, to resist the additional pressure due to the thrust of the roof.

Depth	12		14		16		18		20	
	9	3	9	3	9	3	9	3	9	3
0–4	1	0	1	0	1	0	1	0	1	0
4–8	1	0	1	0	1	0	1	0	1	0
8–12	1	0	1	0	1	0	2	0	2	0
12–16	1	0	2	0	2	0	2	0	2	1
16–20	2	0	2	0	2	1	3	1	3	1
20–24	2	0	2	1	3	1	3	1	3	1
24–28	2	1	3	1	3	1	3	1	0	2
28–32	3	1	3	1	3	1	0	2	0	2
32–36	3	1	3	1	0	2	0	2	0	2
36–40	3	1	0	2	0	2	0	2	0	2
40–44	3	2	0	2	0	2	0	2	0	3
44–48	0	2	0	2	0	2	0	3	0	3
48–50	0	2	0	2	0	2	0	3	0	3

The Iowa Experiment Station advises the use of hard- or high-carbon wire for reënforcement, since it is as cheap as any, is stronger, and does not kink so badly in handling. Since the wire is wound in comparatively small coils, it must

CONSTRUCTION OF FARM BUILDINGS 155

be straightened sufficiently to lie on the wall with approximately the same curvature as that of the silo wall itself. The method which the above-mentioned station has found to be the most convenient may be described as follows: secure or build a reel from which a coil of the wire may be conveniently unwound. Mount this reel upon a plank where it can turn easily, then secure a short piece of gas pipe close to the reel as shown in Figure 71. Through this pipe draw the wire as it uncoils from the reel, the pipe being so placed that its curvature is opposite that of the wire in the coil. The curvature of the pipe

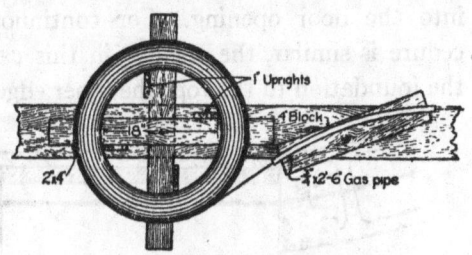

FIG. 71. — Wire straightening device.
(Iowa Agr Exp. Sta)

may be adjusted until exactly the right curvature in the wire is obtained; the wire then may be cut into lengths convenient for handling. The reënforcing wire should be placed on the outer half of the blocks, in order that there be enough mortar inside to bear against the wire and to hold the blocks. The wires should be long enough to lap past each other and admit of the ends being bent back to form a hook, so they may be held together in the form of a hoop.

The roof of this silo may be built of wood, according to the description given heretofore of wood roofs; but since the silo itself is permanent it would seem advisable to make the roof the same, especially so since the cost of a permanent roof is not very much greater than that of a wood roof. The construction of a permanent roof for a silo of this type is similar to that of a hollow-block concrete silo, which is given subsequently in this volume.

The doors for this silo may be either separate or continuous, the latter having been found to be the more convenient. In making provision for an individual door, special forms are provided for which are set in place and surrounded with concrete, the inner part of the resulting opening having a beveled edge against which the door is held. The door is made similar to the one shown in Figure 65, but with a bevel extending all around that just fits into the door opening. For continuous doors the procedure is similar, the opening in this case extending from the foundation to the top, the inner edge of the sides of the

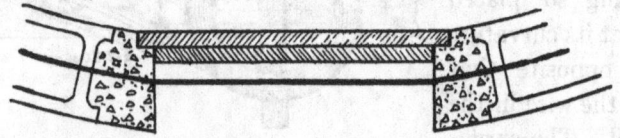

Fig. 72. — Continuous door frame of concrete in a tile silo.

opening having a rabbet an inch square, as shown in Figure 72, against which the door is held; the reënforcement in this case is made of $\frac{5}{8}$-inch stock and is inserted at every third course only, being omitted at the intervening ones, and must be firmly anchored on each side.

The chute for this silo may be built of wood, or preferably of blocks, the blocks being laid up in a single wall not necessarily more than 4 inches thick, and supported at the bottom by some sort of a structure which may serve the additional purpose of housing the silage cart.

Concrete Silos

Concrete silos are made in two ways, either of block or of monolithic construction, in the latter case the whole structure being of one solid mass. The block type of construction has proven popular in colder climates on account of the ease with which a hollow-wall silo can be built, no forms being necessary. Block silos can be built with home labor and of home-made blocks, but where there is a reliable block contractor in the vicinity, it is usually advisable to have the work done by him, on account of not only the saving in time but of numerous other things as well.

The use of concrete as a material for the construction of silos has become almost universal. Enough concrete silos have been built and have been in use a sufficiently long time to prove their unquestionable value. The old idea that the juices of the silage permeated the concrete walls with a deleterious effect upon the latter has been entirely disproved; silos which have been filled with silage for years give no evidence of disintegration or even of discoloration of the interior walls. The fireproof qualities of concrete must also be considered in comparing silos, and instances in which silos have successfully undergone the action of fire which destroyed adjacent or surrounding buildings has shown conclusively that concrete silos are absolutely fire-resistant. The question of permanence is also very satisfactorily settled by the concrete silo, for such a silo properly built and reënforced will stand indefinitely, increasing in strength with age.

There have been cases of the failure of concrete silos, but in almost every case the failure was the result either of poor design or of poor workmanship. In early days the strength of concrete was not very well known, and the reënforcing was insufficient. In some rare instances poor materials were the direct cause of the failure. However, there is no reason why the concrete silo should fail, if the materials are properly selected, if the reënforcement is sufficient and correctly put in, and if the work is done in an approved way.

The cost of a concrete silo compares very favorably with that of other types. The cost will vary greatly in different localities, since local conditions govern the cost to such a great extent. The determining factors are the prices of material and the cost of labor, and where these are at all reasonable, a good concrete silo can be built at a cost no greater than that of a good wood, brick, or tile silo.

The following table gives some excellent data on the cost of 110 concrete silos. In the compilation of the data, material, labor, superintendence, and all miscellaneous expenses incident to putting the structure in shape to be filled, were included.

State	Cost per Ton of Capacity	
	Monolithic	Block
Illinois	$2.83	$2.44
Michigan	2.31	3.21
Wisconsin	2.10	3.36
Minnesota	2.26	3.34
Average cost of all silos, capacity 100 tons or less	2.89	3.52
Average cost of all silos, capacity 100 tons to 200 tons	2.38	2.88
Average cost of all silos, capacity 200 tons or over	2.18	
Average cost of all silos	2.30	3.11

Each of the two methods of concrete construction available for silo work, the monolithic and the concrete block, has certain advantages and disadvantages, but the matter of personal choice generally influences the decision to build either with monolithic walls or with block. The monolithic silo is generally the easier of the two for inexperienced persons to build, and is usually a little cheaper than the block, as it does not require the service of good masons or the use of a block machine: the block silo, however, makes

CONSTRUCTION OF FARM BUILDINGS 159

the use of forms unnecessary, produces a wall with continuous vertical air spaces, and slightly reduces the amount of materials used.

The decision to build either of monolithic or of block construction very often depends upon the availability of materials. In localities where materials for monolithic work are abundant and of good quality, it is hardly practical to haul blocks farther than eight or ten miles; on the other hand, if there is no good sand or gravel near by, block work may be preferred to the monolithic. In such cases, it may be found economical to haul blocks from a greater distance, or make them on the site, if need be.

Foundations

Laying out the Work. — The site of the silo having been selected and its size determined, the excavation should be laid out. This may be done conveniently with a sweep similar to the one shown in Figure 73. A heavy stake is driven in the center of the spot selected for the silo and allowed to project above the surface about 1 foot. The arm of the sweep may be made of a 2 × 4 at least 2 feet longer than half the inside diameter of the silo.

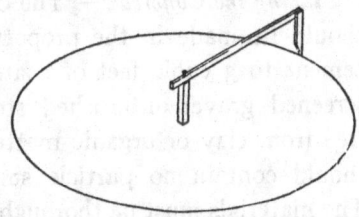

FIG. 73. — A simple sweep for laying out excavation of a silo.

The arm swings about the stake as a center, being held to the latter by a large spike. A chisel-shaped board or template is swung around the stake, and should describe a circle with a diameter 2½ feet greater than the inside diameter of the completed silo. This will give the outline of the excavation and also of the foundation.

Excavating. — The excavation should be carried to a depth not to exceed 6 feet below the floor of the barn where the silage is to be fed. The objection to going deeper is that it adds to the labor in removing the silage. In all cases, however, the foundation should be established below frost. All of the earth within the line described by the sweep should be removed down to a point 1 foot from the bottom, and below this the excavation should be made the shape and size of the foundation, 2 feet wide by 1 foot in depth, so placed that the outer edge will come directly up to the edge of the excavation, assuming that the sides of the latter are perpendicular.

If the silo is to be equipped with a concrete chute, the foundation for the chute should be put in at the same time as that for the silo. As the chute is rectangular in shape, no difficulty should be encountered in excavating for the foundation, which will be at the same depth as the silo foundation, and 2 feet in width by 1 foot in depth.

Placing the Concrete. — The concrete for the foundations should be made in the proportion of 1 sack of Portland cement to 3 cubic feet of coarse sand, to 5 cubic feet of screened gravel or crushed stone. The sand should be free from clay or organic matter, and the gravel or stone should contain no particle smaller in size than $\frac{1}{4}$ inch. The materials must be thoroughly mixed and enough water added to give a quaky consistency. The concrete may usually be placed in the excavation without any forms whatever, but in some kinds of soil light boards, held in position by stakes, may be necessary. The top of the foundation must be leveled off with a straight-edged board and spirit level. After 24 hours, the foundations have generally hardened sufficiently so that the wall may be built upon them. Where soft ground or quicksand is encountered, the foundation may be made 3 or 4 feet in width, to provide plenty of footing.

Imbedding Reënforcing Rods. — If a monolithic silo is to be built, the vertical reënforcing for the walls, consisting of ½-inch round rods spaced 3 feet apart, should be imbedded in the foundation a distance of 8 or 9 inches. If a block silo is to be built, no vertical reënforcing need be placed.

The Floor. — After the foundation is completed, the earth within should be dug out for a depth of about 8 inches, and a concrete floor built as shown in Figure 74. The floor

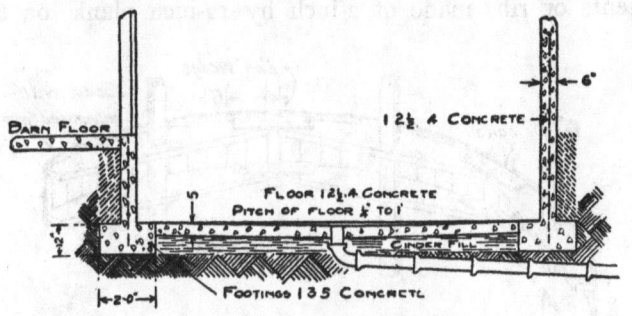

FIG 74. — Silo foundation

should be given a slight pitch in all directions toward the center, and if necessary, an outlet to a line of drain tile should be put in. Outlets are not usually provided in silo floors, but in a few instances silos have failed because of the pressure of a large quantity of water accumulated under unusual conditions, with no provision for escape. In such cases the stress on the walls may reach two or three times that usually imposed by the silage. Although the majority of silos are not provided with a drain, it is undoubtedly a desirable feature. The top of the drain should be protected from accumulations on the silo floor, by a small wire mat. A 4-inch or 6-inch drain tile will be sufficient.

The floor should be made of $1 : 2\frac{1}{2} : 5$ concrete. A smooth finish is not considered necessary.

Monolithic Concrete Silo Walls

Forms. — There are a number of commercial forms on the market for the construction of the walls of monolithic concrete silos. However, homemade forms are perfectly satisfactory when properly made, and a number of excellent silos have been built from such forms.

The homemade forms have the inner part made of segments or ribs made of 2-inch by 12-inch plank, on the

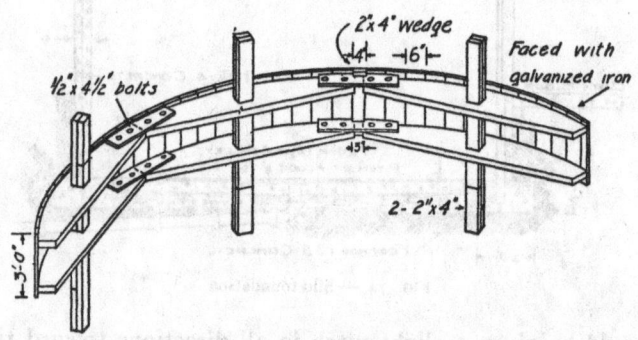

FIG. 75. — Inner wall form.

circumference of which are nailed a number of 1-inch matched floor boards, 3 feet long; this is covered with lightweight galvanized iron. The sections thus made are usually eight in number, all together making a complete circle, and are fastened together when erecting the forms by cleats of 2 × 6 stuff. The outer forms are made of heavy galvanized sheet steel of the same width as the inner forms. Figures 75 and 76 illustrate the details of the construction of these forms. In handling the inner forms great care must be observed in keeping the inside surface

CONSTRUCTION OF FARM BUILDINGS 163

of the silo perfectly smooth. Horizontal "steps" in the wall are particularly objectionable. Projections, "steps," and other irregularities cause uneven settling of the silage, thus forming air pockets. The pressure of an air pocket frequently causes silage within a foot of the pocket on all

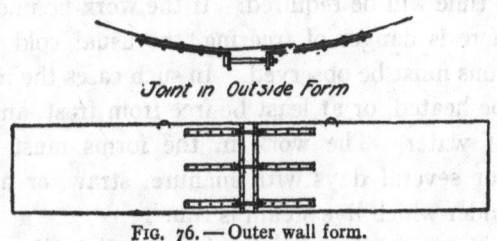

Fig. 76. — Outer wall form.

sides to spoil. The inner surfaces of the forms should be painted, before using, with crude oil or whitewash, which will prevent the concrete from sticking.

Constructing the Walls. — As soon as the foundation has hardened sufficiently to allow the work to proceed, the wall forms may be placed in position. Much care should be taken to locate them centrally and in such a manner that the sides are perpendicular. The 4 × 4-inch uprights should be carefully put in position at this time, being supported on wooden blocks or flat stones. After the inner form is placed, but before the outer form is in position, the horizontal reënforcing rods for the first 3 feet of wall should be wired to the vertical rods which were placed in foundation as previously mentioned. The outer forms should then be placed in position and tightened. Before placing the concrete, it will be necessary to clean off the surface of the foundation and moisten it thoroughly. The wall forms, having been previously painted with crude oil or whitewash to prevent sticking, may then be filled with slushy concrete made in the proportion of 1 sack of

Portland cement to $2\frac{1}{2}$ cubic feet of screened gravel or crushed stone, all of the latter being between $\frac{1}{4}$ inch and $1\frac{1}{2}$ inches in size.

During the summer 24 hours is usually enough for concrete to harden before raising the forms, but in cool weather a longer time will be required. If the work be undertaken while there is danger of freezing, the usual cold weather precautions must be observed. In such cases the materials should be heated, or at least be free from frost, and mixed with hot water. The work in the forms must be protected for several days with manure, straw, or a canvas jacket under which live steam is run.

When the first filling has hardened sufficiently to admit of raising the forms, the forms are raised in position for the next course. Immediately before the concrete is placed for each succeeding course, the surface of that previously laid should be thoroughly cleaned off and moistened, and coated with a cement and water grout of about the consistency of cream. This precaution is necessary to secure a good bond between the courses. It should be observed in all cases, as the pressure of the silage is apt to force moisture through any seams which might occur because of imperfect bond. Concreting should not be discontinued with a course partially completed, but if this is unavoidable, the concrete surface should be left as nearly vertical as possible.

Although the forms are made 3 feet in height, the height of the wall built at each filling (after the first) will be 2 feet 6 inches, allowing the forms to cover 6 inches of finished wall when in position to be filled again. Experiment has shown that this is about the best height to fill at one time, as it makes about one-half day's work for the average farm crew when the mixing is done by hand. In reasonably good weather it should be possible for home labor to raise the form each morning, refill in the forenoon, and have the remainder of the day free for various farm duties.

SPACING OF HORIZONTAL REËNFORCING RODS IN MONOLITHIC SILOS

DEPTH FROM TOP IN FEET	DIAMETER OF SILO IN FEET											
	10 ft				12 ft				14 ft		16 ft		
	DIAMETER OF RODS IN INCHES											
	$\frac{3}{8}$	$\frac{1}{4}$	$\frac{1}{2}$	$\frac{3}{8}$	$\frac{1}{4}$	$\frac{1}{2}$	$\frac{3}{8}$	$\frac{1}{4}$	$\frac{3}{8}$	$\frac{1}{2}$	$\frac{3}{8}$	$\frac{1}{4}$
From to						Inches						
0–4	—	18	—	—	18	—	—	18	—	—	—	14
4–8	—	18	—	—	17	—	—	15	—	—	—	11
8–12	—	15	—	—	12	—	—	10	—	—	—	9
12–16	—	11	—	—	9	—	17	$7\frac{1}{2}$	—	—	15	7
16–20	—	9	—	16	7	—	13	6	—	—	12	6
20–24	16	7	—	13	6	—	12	5	—	—	11	5
24–28	14	6	—	12	5	18	10	5	—	15	9	4
28–32	12	6	18	10	5	15	9	4	—	14	8	4
32–36	11	5	16	9	4	14	8	4	—	12	7	3
36–40	10	4	14	8	4	12	7	3	17	11	6	3
40–44	9	4	13	7	3	12	7	3	15	10	6	3
44–48	—	—	—	—	—	10	6	3	14	9	5	—
48–50	—	—	—	—	—	—	—	—	—	—	—	—

DEPTH FROM TOP IN FEET	DIAMETER OF SILO IN FEET											
	18 ft				20 ft					22 ft			
	DIAMETER OF RODS IN INCHES											
	$\frac{5}{8}$	$\frac{1}{2}$	$\frac{3}{8}$	$\frac{1}{4}$	$\frac{3}{4}$	$\frac{5}{8}$	$\frac{1}{2}$	$\frac{3}{8}$	$\frac{1}{4}$	$\frac{3}{4}$	$\frac{5}{8}$	$\frac{1}{2}$	$\frac{3}{8}$	$\frac{1}{4}$
From to														
0–4	—	—	—	12	—	—	—	—	11	—	—	—	—	10
4–8	—	—	—	10	—	—	—	—	9	—	—	—	—	9
8–12	—	—	—	8	—	—	—	16	7	—	—	—	15	$6\frac{1}{2}$
12–16	—	—	$13\frac{1}{2}$	6	—	—	—	12	$5\frac{1}{2}$	—	—	—	11	5
16–20	—	—	11	5	—	—	18	10	5	—	—	16	9	4
20–24	—	18	10	5	—	—	16	9	4	—	—	15	8	4
24–28	—	14	8	4	—	—	13	7	3	—	—	11	7	3
28–32	—	12	7	3	—	17	11	6	3	—	16	10	6	3
32–36	17	11	6	3	—	15	10	5	3	—	14	9	5	3
36–40	15	10	6	3	—	14	9	5	—	—	12	8	5	—
40–44	14	9	5	—	—	12	8	5	—	16	11	7	4	—
44–48	13	8	5	—	16	11	7	4	—	15	10	7	4	—
48–50	12	7	4	—	15	11	7	4	—	14	10	6	3	—

Reënforcing. — Steel rods are preferable to other kinds of reënforcing because they come in standard sizes, the strength of which is definitely known. For all silos, regardless of diameter or height, the vertical reënforcing should be ½-inch round or twisted rods placed in the middle of the wall at intervals of about 3 feet. The size and spacing of the horizontal reënforcing depend upon the diameter of the silo and the distance from the top. The first horizontal rods should be placed 2 inches above the foundation. Wherever rods are spliced, they must be lapped for a distance equal to 64 times the diameter, which is 16 inches for ¼-inch rods, 24 inches for ⅜-inch rods, and 32 inches for ½-inch rods. Immediately before the outer form is raised to position, the horizontal rods should be wired in place for a distance equal to the height of the forms. Where a concrete cornice is put on, an extra reënforcing band is put around the top for the purpose of strengthening it.

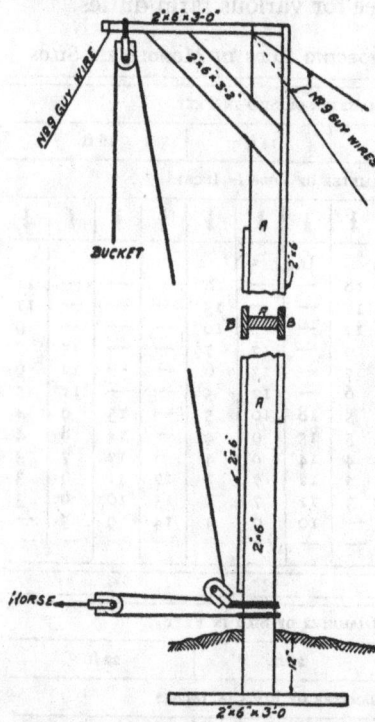

FIG. 77. — A hoisting derrick for concrete silos.

The work of constructing the silo will be made much easier if a convenient method of hoisting materials is adopted at the start. The old scheme of raising the concrete by hand with a rope and bucket wastes time and materials, besides incurring unnecessary and disagreeable labor. Materials may best be raised with a rope and pulley, the latter attached to a derrick frame, the construction of one such frame designed by the Iowa Experiment Station being illustrated in Figure 77; this has been tested and found safe for loads not exceeding 400 pounds.

A few monolithic concrete silos have been constructed with hollow walls, and have proved to be eminently satisfactory, though somewhat high in cost. Their chief advantage lies in the presence of a continuous air chamber surrounding the silo, which acts as an insulator, so the trouble from freezing is reduced to a minimum.

Concrete Block Silos

When the work is done by a contractor, the owner should take the precaution of examining the blocks which go into his silo, rejecting those that are damaged or of inferior quality. A crack of any size, or broken or crumbly edges, indicate a weakness in the block and make it unsuited for use. Blocks may be tested for their water-resisting qualities by placing a small amount of water on the surface and observing whether this remains or is absorbed. A block which readily absorbs moisture is obviously unsuited for silo work, which dampness must not penetrate. Warped and distorted blocks should be discarded because of their unsightly appearance.

Laying the Blocks. — The foundation already described will give as good satisfaction for the block silo as for the monolithic. The top of the footing must be made perfectly level, being tested frequently with a level board. As soon as the footing has sufficiently hardened, the top should then be cleaned off and moistened and a coat of slushy mortar ¼ inch thick put on. The first band of reënforcing should then be put in, and the first row of blocks laid on this mortar, beginning the blocks at the two ends of the wall next to the doorway and continuing around. The blocks may be more conveniently set in a true circle if a sweep similar to the one used in laying out the foundation is used here. Should the blocks fail to meet exactly, the circle should be enlarged or made a little smaller, whichever happens to be the more convenient. A guide board with a convex edge, cut on a circle of the same diameter as the inside of the silo, should then be made and used in place of, or in conjunction with, the sweep in laying up the remaining courses.

The Mortar. — The mortar should consist of 1 sack of Portland cement to 2 cubic feet of coarse sand, with the possible addition of a small quantity of lime (not over 10 per cent), if need be, to make it easier to work. Before laying up the blocks see that they are thoroughly sprinkled, which will prevent them from drawing moisture from the mortar. No more mortar should be mixed at one time than can be used up within 30 minutes after the first moistening. If lime is used, it must be thoroughly slaked.

Reënforcing. — The only failures reported on block silos have been due to a lack of sufficient reënforcing, caused in most cases by the overconfidence of the builder in the strength of the blocks, or failure to realize the enormous outward pressure of the silage. Horizontal reënforcing is of the utmost importance and must not be overlooked. Vertical reënforcing in block silos is not considered necessary. The accompanying table shows the size of rod which should be placed between each row of blocks or in the groove in each row of blocks, if such a groove is provided. Reënforcing rods in block silos are not lapped in the ordinary fashion, but are anchored around a block, or the ends are hooked together.

HORIZONTAL REËNFORCEMENT FOR BLOCK SILOS

Showing size Wire and Rods to be used between each course of blocks 8 inches high

Feet from Top of Silo	DIAMETER OF SILO							
	8 ft.	10 ft.	12 ft	14 ft	16 ft	18 ft	20 ft.	22 ft
0–4	No. 6	No. 6	No. 6	No. 6	No. 6	$\tfrac{1}{4}''$	$\tfrac{1}{4}''$	$\tfrac{1}{4}''$
4–8	No. 6	No. 6	No. 6	No. 6	No. 6	$\tfrac{1}{4}''$	$\tfrac{1}{4}''$	$\tfrac{1}{4}''$
8–12	No. 6	No. 6	$\tfrac{1}{4}''$	$\tfrac{1}{4}''$	$\tfrac{1}{4}''$	$\tfrac{1}{4}''$	$\tfrac{3}{8}''$	$\tfrac{3}{8}''$
12–16	No. 6	$\tfrac{1}{4}''$	$\tfrac{1}{4}''$	$\tfrac{1}{4}''$	$\tfrac{1}{4}''$	$\tfrac{3}{8}''$	$\tfrac{3}{8}''$	$\tfrac{3}{8}''$
16–20	No. 6	$\tfrac{1}{4}''$	$\tfrac{3}{8}''$	$\tfrac{3}{8}''$	$\tfrac{3}{8}''$	$\tfrac{3}{8}''$	$\tfrac{3}{8}''$	$\tfrac{3}{8}''$
20–24	No. 6	$\tfrac{3}{8}''$	$\tfrac{3}{8}''$	$\tfrac{3}{8}''$	$\tfrac{3}{8}''$	$\tfrac{3}{8}''$	$\tfrac{3}{8}''$	$\tfrac{3}{8}''$
24–28	$\tfrac{1}{4}''$	$\tfrac{3}{8}''$	$\tfrac{3}{8}''$	$\tfrac{3}{8}''$	$\tfrac{3}{8}''$	$\tfrac{3}{8}''$	$\tfrac{1}{2}''$	$\tfrac{1}{2}''$
28–32	$\tfrac{1}{4}''$	$\tfrac{3}{8}''$	$\tfrac{3}{8}''$	$\tfrac{3}{8}''$	$\tfrac{3}{8}''$	$\tfrac{1}{2}''$	$\tfrac{1}{2}''$	$\tfrac{1}{2}''$
32–36	$\tfrac{1}{4}''$	$\tfrac{3}{8}''$	$\tfrac{3}{8}''$	$\tfrac{3}{8}''$	$\tfrac{1}{2}''$	$\tfrac{1}{2}''$	$\tfrac{1}{2}''$	$\tfrac{1}{2}''$
36–40	$\tfrac{1}{4}''$	$\tfrac{3}{8}''$	$\tfrac{3}{8}''$	$\tfrac{1}{2}''$	$\tfrac{1}{2}''$	$\tfrac{1}{2}''$	$\tfrac{1}{2}''$	$\tfrac{1}{2}''$
40–44	$\tfrac{1}{4}''$	$\tfrac{3}{8}''$	$\tfrac{3}{8}''$	$\tfrac{1}{2}''$	$\tfrac{1}{2}''$	$\tfrac{1}{2}''$	$\tfrac{1}{2}''$	$\tfrac{1}{2}''$
44–48	$\tfrac{3}{8}''$	$\tfrac{3}{8}''$	$\tfrac{1}{2}''$	$\tfrac{1}{2}''$	$\tfrac{1}{2}''$	$\tfrac{1}{2}''$	$\tfrac{1}{2}''$	$\tfrac{3}{8}''$
48–50	$\tfrac{3}{8}''$	$\tfrac{3}{8}''$	$\tfrac{1}{2}''$	$\tfrac{1}{2}''$	$\tfrac{1}{2}''$	$\tfrac{1}{2}''$	$\tfrac{1}{2}''$	$\tfrac{5}{8}''$

For illustration, let it be assumed that the proper method of reënforcing a silo 32 feet in height and 16 feet in diameter is desired, blocks 8 inches in height being used. Referring to the table, we run down the vertical column at the left until the figures indicating the greatest depth of the silo are reached. In this case these figures are 28–32 feet. Running directly across horizontally to the 16-foot diameter column, we find that the proper reënforcing 28–32 feet from the top of the silo is one $\tfrac{3}{8}$-inch rod between each course of blocks; following up directly the 16-foot diameter column, we find that $\tfrac{3}{8}$-inch rods must be used between each course until a point 16 feet from the top is reached. From here up $\tfrac{1}{4}$-inch rods are used until 8 feet from the top when No. 6 rods are substituted.

The reënforcing is commonly laid in the mortar between

the courses of blocks, the strength of the mortar and the downward pressure of the blocks above being depended upon to keep the rods in place under loaded conditions. In the best practice, however, blocks are used which have a recess in the top face deep enough to accommodate the reënforcing rod. Recesses are generally put about 2 inches from the outside of the block.

Doorways

Continuous and noncontinuous doorways are used about equally in monolithic silo construction, and the question of which to use is generally settled by personal choice. The continuous doorway has the advantage of providing a

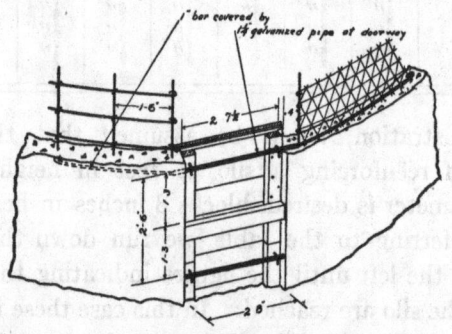

FIG. 78. — Continuous doorway for monolithic silo.

larger space through which to throw the silage, and for this reason is preferred by many. The noncontinuous doorways, as used by some of the best contractors, have no disadvantage except that they provide a smaller space through which to remove the silage.

Continuous Doorways for Monolithic Silos. — A satisfactory continuous doorway can be made by forming concrete

CONSTRUCTION OF FARM BUILDINGS   171

jambs on both sides of the opening. This is easily accomplished by inserting between the forms, at proper distances apart, vertical wooden forms to mold the face of the jamb and the recesses into which the doors will fit. Where the concrete chute is built simultaneously with the silo walls, the vertical jamb forms will extend from the inner wall to the inner chute form. If

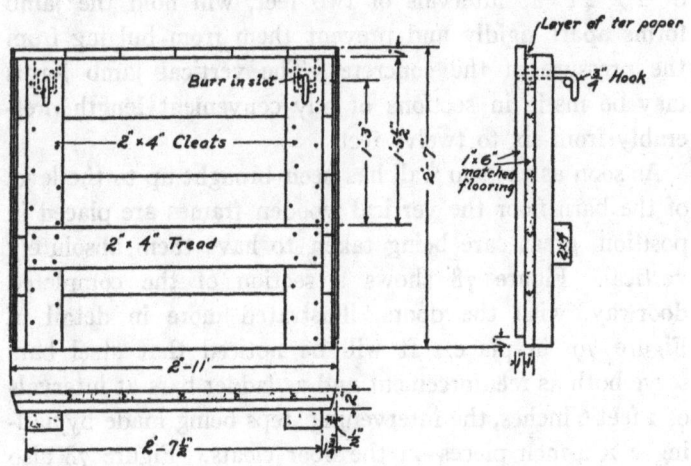

FIG. 79. — One method of making a door for a continuous doorway.

the silo walls are constructed without the chute, the jamb forms must be placed between the inner and the outer wall forms.

The forms for casting the face of the jambs may consist of 2-inch planks of a width equal to the distance between the wall forms. Strips of 2 × 2-inch material should be nailed to the face of the planks so as to form 2 × 2-inch vertical recesses on the inside of the opening. Horizontal slots, to accommodate the ladder rounds, will have to be made in the planks at intervals of 18 inches. All surfaces of wood which will come into contact with the concrete should be planed and oiled, which will insure a smooth surface and prevent the wood from adhering to the concrete.

The distance between face of the jambs should be 30 inches and the jamb forms rigidly maintained in a vertical position and at proper distance apart. Spacers consisting of 2 × 4's, at intervals of two feet, will hold the jamb forms apart rigidly and prevent them from bulging from the pressure of the concrete. The vertical jamb forms may be made in sections of any convenient length, preferably from six to twelve feet.

As soon as the silo wall has been brought up to the level of the barn floor the vertical wooden frames are placed in position, great care being taken to have them absolutely vertical. Figure 78 shows a section of the completed doorway, with the doors, illustrated more in detail in Figure 79, in place. It will be noticed that steel bars serve both as reënforcement and as ladder bars at intervals of 2 feet 6 inches, the intervening steps being made by nailing 2 × 4-inch pieces on the door cleats. Figure 78 also shows two methods of reënforcing which are used to strengthen the walls of monolithic silos. At the left is shown the ordinary method of employing vertical and horizontal bars firmly fastened together at the intersections. At the right is shown a somewhat simpler, and for smaller silos, quite as satisfactory a method, in which the bulk of the reënforcing consists of a rather heavy closely woven wire cloth, supported at intervals by vertical rods.

Continuous Doorways for Concrete Block Silos. — Concrete jambs for the continuous doorways of concrete block may be made as shown in Figure 80, and faces of the jambs

CONSTRUCTION OF FARM BUILDINGS 173

should be the same as those on the continuous door jambs of monolithic silos, described previously. The jambs may be easily constructed by the use of simple box molds, recesses being formed on the inside of the jambs by the use of 2 × 2-inch cleats. As the reënforcing rods are

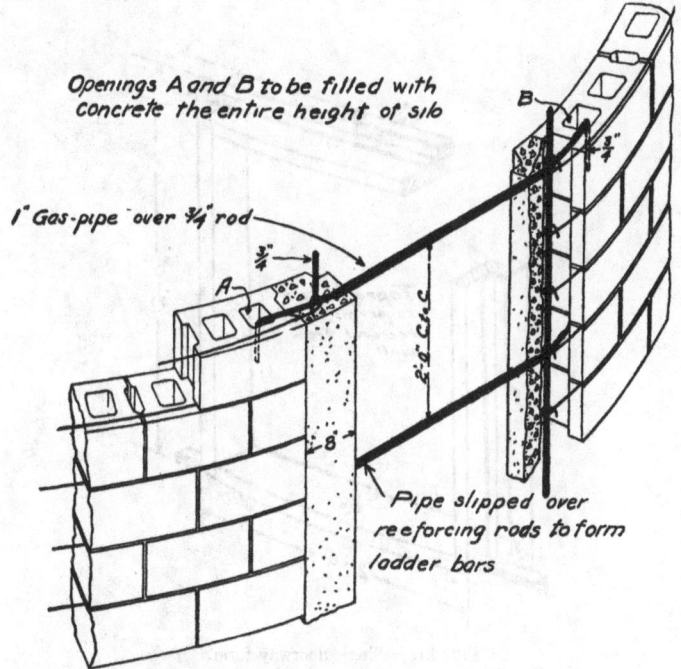

FIG. 80. — Continuous doorway for block silo.

laid upon successive courses of blocks, they are cut off so that the ends will extend out far enough to be firmly fastened to the ½-inch vertical rods to which the horizontal ladder rods are attached. These vertical rods should be located near the center of the jamb. The doors for the continuous doorways of either monolithic or concrete block silos are made either as shown by diagram in Figure 79, or as described on page 175 and illustrated in Figure 82.

Noncontinuous doors. — Noncontinuous doors are perhaps easier to build than continuous doorways, and if the owners are satisfied that they provide sufficient room for

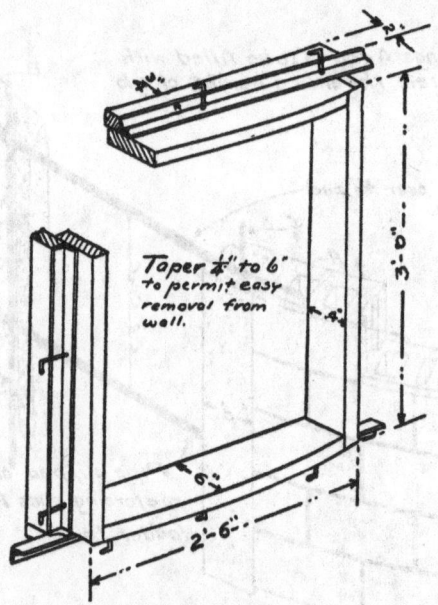

FIG. 81. — Single doorway form.

getting the silage out conveniently, there is no objection to their use, although, on the other hand, they possess no great advantage over doors of the continuous type. The arguments often heard that the noncontinuous-door silo is a stronger type than the other, and *vice versa*, carry little weight, as either type may be made sufficiently strong.

CONSTRUCTION OF FARM BUILDINGS

Noncontinuous doors are often put in with a distance of about 2½ feet between them, but the spacing may vary to suit the individual owner. In all cases the arches between the doors must contain an amount of reënforcing equivalent to the full amount of horizontal reënforcing put around the silo. Thus, if the doors are 3 feet in height, with a distance of 2½ feet between them, the horizontal reënforcing in the space between the doors should be equivalent in amount to that placed in 5½ feet of the wall where there are no doors.

Doorway Form and Frame. — Figure 81 shows a form for a noncontinuous door-opening in a monolithic silo. The bottom and top pieces are made of 2 × 6-inch plank cut to the arc of a circle with diameter the same as the outer diameter of the silo wall. The two sides are made of 2 × 4's. A frame of lighter materials is placed around the outside of the form for the purpose of making a recess 2 inches deep around the opening on the inner side of the wall, into which the door will fit. This

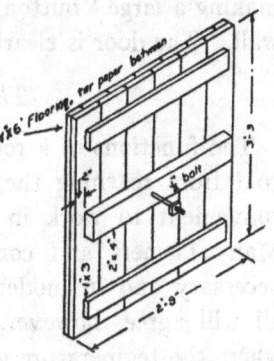

FIG. 82. — Individual door.

frame is tapered to permit removal from the wall as soon as the concrete has hardened. It may then be used again for the next doorway above.

If desired, a door frame of small angle iron (as shown) may be used to protect the corners of the concrete. The frame should be slipped on over the form, and both frame and form then placed in position. The angle iron should be cut a few inches longer than the dimension of the opening and the ends imbedded in the concrete. The frame should also be anchored to the concrete by large spikes. Holes to receive the spikes should be drilled in the angles, 12 inches apart. The spikes should be bent at right angles to secure a better hold in the wall.

Doors. — The doors may best be made of two thicknesses of 1 × 6-inch matched flooring with a layer of tar paper between. The 1 × 6-inch boards are held together by two 1 × 4-inch cleats across the top and bottom and one 2 × 4-inch cleat across the center. The middle cleat is made larger than the others in order to take care of the strain caused by the large bolt in the center. A 2 × 4, 4 inches long, or a similar piece of material, is placed on the bolt, making a large "button," by which the door is held in the wall. The door is clearly shown in Figure 82.

The Concrete Roof

The functions of a roof on a silo are (1) to prevent the cold from reaching the silage, and (2) to make it more convenient to work in the silo during stormy weather. Many farmers and contractors do not consider a roof necessary and in moderate climates this is probably so; all will agree, however, that in sections of the country where the temperature goes below zero a roof is a positive necessity, as well as a great convenience under any circumstances.

The logical way to finish up a permanent silo is with a permanent roof. The tendency at the present time is toward the permanent silo, from foundation to pinnacle. If the directions given in the following paragraphs are closely followed, little difficulty will be found in putting on a permanent roof, one that will last indefinitely without need of being shingled or otherwise repaired, and which will be in no danger of blowing off.

The Cornice. — A cornice is only necessary where a roof is to be put on, its chief uses being to prevent water from the roof from running down the walls, and to improve the appearance of the silo. Figure 83 illustrates how the forms are made for the cornice on a monolithic silo.

The brackets for the forms are made of $\frac{1}{4} \times 2$ inch strap iron bent as shown, and drilled to receive stove bolts. These brackets should be placed on the outer form at intervals of about 6 feet, holes being drilled at the proper points to receive the stove bolts. The bottom of the cornice mold box is made of 2×6 inch planks in short lengths sawed to the arc of a circle with diameter 1 foot larger than that of the inside of the silo. The side of the mold is made of 1×6 inch planks spiked to the bottom boards. The mold is held in place by screws through the bracket, as shown. An extra band of horizontal reënforcing is put in the cornice, as may be seen in the figure. The vertical rods in the silo walls and the radial rods of the roof are all brought around the horizontal reënforcing in the cornice, thus holding it in place and strengthening the cornice.

For the top section of the wall (last filling of the forms) the inner and outer forms are brought up to the line of the

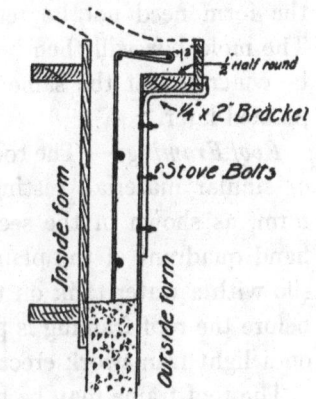
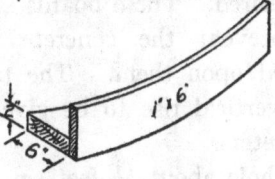

Fig. 83. — Cornice mold box.

top of the completed wall. The forms are then filled to within one foot of the top, the outer form removed, and brackets attached. (If the stove bolts are already in place, the form need not be removed to attach the brackets.) The mold box will then be put in place. The cornice will be concreted at the same time as the roof, as will be explained later.

Roof Framing. — The roof framing may consist of 2 × 4's or similar material, resting on the top of the inner wall form, as shown in the sectional view, and the lower left-hand quadrant of the plan view, Figure 84. In case of a silo with a water tank on top, the forms must be removed before the roof framing is put up, and the latter supported on a light framework erected within the tank.

The roof frame may be boarded up as shown in the plan view, with boards running either radially or otherwise, as desired. These boards should be placed close together to prevent the concrete from coming through when placed upon them. The table given on page 181 shows the vertical rise to be given to roofs for silos of various diameters.

A hole about $2\frac{1}{2}$ feet square must of course be left for filling the silo, or if the roof covers the tank, the hole will afford access to the latter. Before placing the reënforcing or the concrete, the top of the framing should be covered with old newspaper, building paper, or similar material, which will prevent the concrete from sticking to the forms. This will greatly facilitate their removal.

Placing the Reënforcing. — The lower right-hand quadrant of the plan and the sectional view shows the spacing of the radial and hoop reënforcing. The former is placed so that the distance between the three bottom hoops is 6 inches, between the next three hoops 9 inches, and between all

CONSTRUCTION OF FARM BUILDINGS 179

remaining hoops 12 inches. Extra rods should be put in around the window opening if the regular rods do not

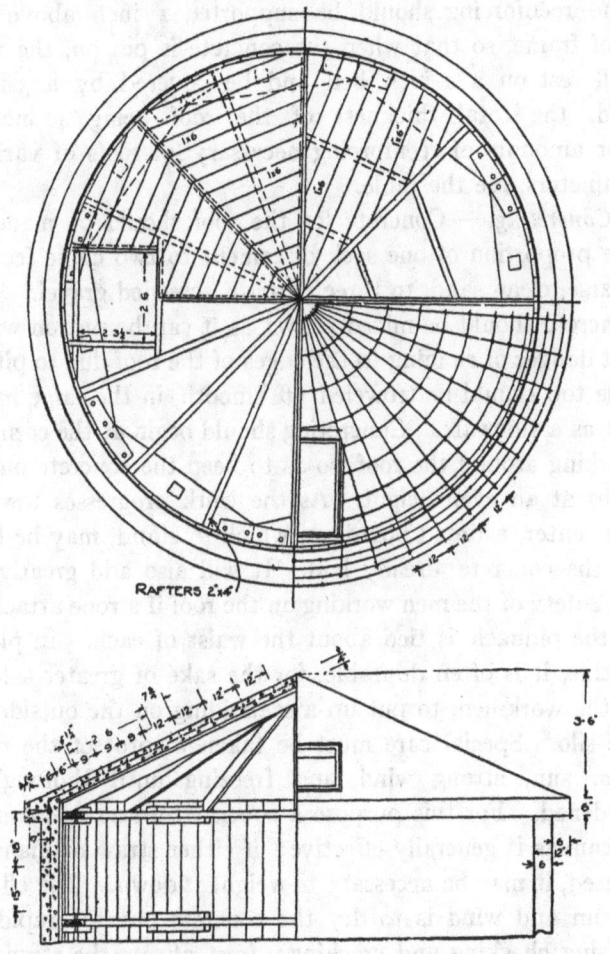

Fig. 84. — Reenforced concrete roof — strong, serviceable, and permanent.

follow the outline of the window closely enough to reënforce it. All intersections must be wired together, and the outer ends of the radial wires brought down and bent around the horizontal reënforcing in the cornice, as shown. The reënforcing should be supported 1 inch above the roof frame, so that when the concrete is put on, the rods will rest on a 1-inch bed and be covered by a 3-inch bed, the total thickness of the roof being 4 inches. For amounts of reënforcing necessary for roofs of various diameters, see the table.

Concreting. — Concrete for the roof should be made in the proportion of one sack of cement to two cubic feet of coarse, clean sand, to three parts of screened gravel. The concrete should be mixed as wet as it can be put on without danger of running to the edges of the roof due to pitch. The top should be troweled off smooth, in the same manner as a sidewalk. Concreting should begin at the cornice, working around the roof, so as to keep the concrete on all sides at an even height. As the work progresses toward the center, a broad board, on which to stand, may be laid on the concrete already laid. It will also add greatly to the safety of the men working on the roof if a rope attached to the pinnacle is tied about the waist of each. In place of this, it is often desirable, for the sake of greater safety to the workmen, to put up a scaffolding on the outside of the silo. Special care must be taken to protect the roof from sun, strong wind, and freezing until thoroughly hardened. For this purpose a covering of straw, manure, or canvas is generally effective; if either straw or manure is used, it may be necessary to weight it down. The effect of sun and wind is to dry the concrete out too rapidly, causing checking and cracking; frost affects the strength of the concrete and is otherwise objectionable.

Monolithic Roofs for Hollow Block Silos. — Where it is desired to put a monolithic concrete roof on a hollow block

CONSTRUCTION OF FARM BUILDINGS 181

silo, the wall should be laid up in the usual manner until the third course of block from the top is reached. The blocks used in this course should be solid, that is, made without cores, or if with the cores, these should be filled up with mortar. The last two courses of hollow block should then be laid, the cores being left open.

DIMENSIONS AND MATERIALS FOR ROOFS
FOR SILOS WITH DIAMETERS 8 FEET TO 22 FEET

Diameter of Silo	Vertical Rise	Volume of concrete in cu. yds.	Cement required bbls	Sand required cu yds	Stone required cu yds	$\frac{1}{4}''$ REËNFORCING RODS		
						No of rods req'rd	Stock length of rods	No of pounds of rods
8 ft.	2 ft.	0.63	1.09	0.33	0.49	26	10 ft.	42
10 ft.	2½ ft.	1.01	1.75	0.52	0.78	31	12 ft.	62
12 ft.	3 ft.	1.49	2.59	0.77	1.15	33	16 ft.	88
14 ft.	3½ ft.	2.05	3.56	1.07	1.58	45	16 ft.	120
16 ft.	4 ft.	2.71	4.72	1.41	2.08	87	10 ft.	146
18 ft.	4 ft.	3.34	5.80	1.74	2.57	93	12 ft.	187
20 ft.	4 ft.	4.11	7.15	2.13	3.17	107	12 ft.	226
22 ft.	4 ft.	4.93	8.55	2.56	3.80	113	14 ft.	265

Concrete for roofs is made of 1 sack of Portland cement to 2 cubic feet of coarse sand to 3 cubic feet of stone. Each cubic yard of concrete requires 1¾ barrels of cement, ½ cubic yard of sand, and ¾ cubic yard of stone, approximately. The ¼-inch reënforcing rods weigh 16.7 pounds per 100 feet.

Special cornice blocks should be cast to make the cornice projection. The block should be 14 inches in width and of the same length on the inside of the wall as the wall blocks. The portion of the cornice blocks directly above the wall blocks should be 6 inches thick, so as to give a 1-inch drop. The roof framing is then put up in the same manner as described before, but in this case it must be supported by the scaffolding instead of on the inner form mentioned there. The reënforcing is placed in the same manner as described before, excepting that the outer ends of the radial rods are made to extend down through the holes in the block for a distance of a foot or more. Since the holes in the third course of block from the top were either omitted or filled up before these blocks were laid, holes in the upper courses can be filled up with wet concrete as soon as the reënforcing rods are in position. The roof is concreted as described previously. Before the concrete is placed on the cornice blocks the latter must be moistened and painted with a cement and water grout.

Concrete Chutes

A permanent chute of concrete is a valuable adjunct to any concrete or masonry silo. The same arguments presented for the concrete silo stand for the chute. The concrete chute is substantial and permanent, fireproof and coldproof, and it greatly improves the appearance of the silo.

Size of Chute. — Chutes in use in various parts of the country vary in size from 2 feet square to about 5 feet square (inside dimensions), but the former size is much too small and the latter larger than need be. For the average monolithic silo a chute 3 feet by 4 feet in inside dimensions is recommended. The outer dimensions will then be 4 feet by $4\frac{1}{2}$ feet, the walls being 6 inches thick. A monolithic chute of this size will require $\frac{1}{3}$ of a barrel of cement, $\frac{1}{8}$ cubic yard of sand, and $\frac{1}{5}$ cubic yard of gravel, per foot of height. For the block silo, the size should be such as will be accommodated by whole and half blocks. The outer dimensions of a hollow-block chute (using $8 \times 8 \times 15$-inch blocks) should be 4 feet 8 inches square, making the inside dimensions 3 feet 4 inches

CONSTRUCTION OF FARM BUILDINGS   183

by 4 feet. This size will require 9½ blocks for each course.

Foundations. — The foundation for the chute should be 2 feet wide and 1 foot high, the same as that for the silo, using concrete of the same proportions. If a monolithic chute is to be built, ⅜-inch vertical reënforcing rods must be imbedded in the foundation 18 inches apart. Monolithic chute walls may be built up simultaneously with the silo walls, but it is much more convenient to build them after the completion of the latter; chute walls of concrete block must be built at the same time, being built in and kept at the same level as the silo walls.

Monolithic chutes. — The accompanying illustration, Figure 85, shows forms in position for building a monolithic chute. Two-inch planed lumber should be used for the face of the forms, and 2 × 4's for the vertical braces. The steel rods used to hold the forms together should be 24 inches long, threaded for 4 inches at each end. Each section of the form should be about 2 feet high. To raise the forms the lower rods are withdrawn and the holes made by them cemented up. The wooden braces are then raised, and the lower panels of planks placed above the others.

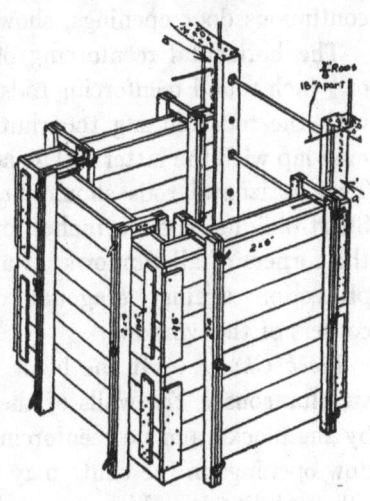

FIG. 85 — Concrete chute forms.

The method of joining the chute to the silo is shown in the figure. Two 1 × 6-inch boards, with edges slightly beveled to permit of easy removal, are placed in a vertical position on the inside of the outer silo form, 3 inches to each side of the line of the doors. In this manner recesses *a* are produced. Three-eighths-inch rods 30 inches long, spaced at intervals of 18 inches, and bent as shown by the dotted lines in the figure, are used to hold the chute securely to the silo. The most convenient way to put in these rods is to have them slightly stapled to the boards occupying recesses *a*. This will hold the rods in position until the concrete is placed. The forms and vertical boards may be removed as soon as the walls have hardened sufficiently, and the ends of the rods bent up into a horizontal position. Where windows are desired in the chute, the openings may be made with a form similar to that used for making non-continuous door openings, shown in Figure 80.

The horizontal reënforcing of the chute should consist of $\frac{3}{4}$-inch round reënforcing rods so spaced as to correspond with the rods binding the chute to the silo, so that they may lap with the latter. The lap should be 24 inches long. Two horizontal rods should be placed over all windows. Short oblique rods, 24 inches long, should be put in about the corners of all windows, at an angle of 45 degrees, as a protection against diagonal cracks running from the corners of the windows.

Block Chutes. — If the block silo and chute are put up simultaneously, the walls of the two will be held together by the blocks, and no reënforcing will be necessary. Window openings in the chute may be made by using concrete sills and lintels, which are easily obtainable from block dealers. A length of heavy strap iron may be substituted for the lintel, if desired, and the sill cast in place by means of a simple box mold.

The Pit Silo

Two very important considerations, economy and durability, are combined in a type of silo known as the "pit silo," which may be wholly underground, or partly above ground. This type of silo is a return to the original form, that of a hole in the ground, and is being used in certain localities with some degree of success.

The simplest method of construction is as follows: excavation is begun by digging a trench a foot wide and a foot deep around the top, with the inner diameter the same as that of the pit; this trench is filled with concrete, and when the concrete is hardened, the earth within is removed to a depth of five or six feet. The earth wall is then covered with a cement plaster, several coats being applied to make a total thickness of an inch. When this portion of the wall is completed, another five or six feet of earth is removed, and the walls covered with plaster. This process is repeated until the required depth is reached, when a floor may be put in, if desired. If part of the silo is built above ground, the concrete ring at the top should be made heavier, in order that it may serve as a foundation. Should the soil be loose and show a tendency to cave in, it may be not only desirable, but necessary, to construct a four-inch wall of concrete or of brick laid in cement.

The chief advantage of the pit silo lies in its cheapness and simplicity of construction. It has serious disadvantages, however, inasmuch as a perfectly drained site is necessary for its successful construction, and as some special contrivance must be designed for hoisting the silage out of the pit; this is likely to prove difficult and expensive. The accumulation of carbon dioxide in the bottom must also be looked out for, for this gas is given off by silage and may collect in sufficient density to asphyxiate a person working in the silo.

POULTRY HOUSES

There is no subject connected with poultry production as important as the housing; not only the comfort, but the health and the productiveness of the fowls depend largely upon proper housing. The house that fulfills all ideal conditions has not yet been constructed; the best of them have their defects. The open front, in the continuous or long house, and the open-front colony type are rapidly making headway, and by most progressive poultrymen are considered the best type. The open-front house, with certain modifications, is used successfully even in Canada, in regions where the temperature falls to 40 degrees below zero.

The widespread interest in the housing of poultry has resulted in a marked improvement in poultry-house construction, though, as mentioned before, the best type of house is yet to be built. The essential requirements of a good poultry house are: good location, dryness, ventilation, sunlight, convenience, ease of disinfection, economy in construction, and, for certain conditions, portability. Much has been said along these lines, but the lack of definite, economic information is felt; it certainly is not a good commercial proposition to invest $5 per fowl in houses alone, if $1 will accomplish the purpose as advantageously.

The poultry house should be located on a site in which the drainage is especially good, if possible, on light, porous, sandy soil sloping gently to the south. Should such a location not be available, the best site possible should be selected, and artificial drainage beneath the floor of the house and for the surrounding soil must be provided. Good air drainage is essential, and for this reason the house should be located at the top of the slope.

The location of the house with reference to the rest of the buildings of the farm is not a small consideration. Placing the house in close proximity to the other buildings has been objected to because the hens are inclined to overrun and inhabit them, thus becoming a nuisance. However, this can be obviated by fencing in the poultry yard. On the farm the women usually care for the poultry, and their work should not be increased by a trip of several hundred yards to the poultry house several times a day.

Foundations. — A stationary poultry house should have a good foundation, one that is substantial and verminproof. Concrete satisfies these requirements most efficiently. The concrete foundation wall need not be especially heavy, a wall 6 inches in thickness with a footing 10 inches in width, the whole extending below the frost line, being amply sufficient. The wall should extend at least 8 inches above the ground line to protect the lower part of the superstructure from rot. Bolts must be inserted into the concrete to which to fasten the sills, for a light structure such as a poultry house is likely to be blown over if not well anchored.

Floors. — Three types of floors have been used in poultry-house construction; namely, earth, wood, and concrete. The first is of course the cheapest, and by some authorities is claimed to be the best, since a dust bath is always available. However, it is easily contaminated by diseases, is hard to keep clean and fresh, and unless exceptionally well drained, is always damp. If the earth floor must be used, 4 or 6 inches of the earth at the surface should be removed each year and replaced with fresh earth. This should be occasionally spaded up and sprinkled with lime as a disinfectant.

The board floor should be used only in colony houses where the required portability would preclude the use of any other type. These are rather expensive, not permanent, and furnish excellent quarters for harboring vermin.

Concrete floors are increasing rapidly in use and popularity. Since they are not subjected to any severe use, they can be built rather thin, 3 inches in thickness making a floor amply strong. The first cost of concrete floors is greater than that of other types, but the labor they save will soon pay for this. They are durable, dry, clean, and in the event of disease can be easily and completely disinfected. In order to insure dryness, the finish coat should be of an inch thickness of cement and sand, in the proportion of 1 part of cement to 2 of sand, the whole mixture well waterproofed. The concrete floors should be given a coat of hot asphaltum, both as a moisture preventive and a protection for the claws of the fowls. It often happens that with the constant scratching on the concrete floor during the winter months the fowls will wear the toes down to the quick until they bleed, and this can be avoided by the use of an asphaltum coat on the floor.

Ventilation. — For a long time there pervaded the realm of poultry fanciers the idea that fowls, in order to thrive, must be tenderly housed in the wintertime, and kept warm and comfortable in a close house. Years of experiment with heated houses, then with glass-front or hothouse construction, seemed to prove that these were incorrect, not meeting the needs of the fowls, as indicated by their decreasing vitality, the low egg production, and the large number of sick and dead during the year. The next step in poultry-house construction was a radical one, the change being made from the closed warm house to the open or curtain-front type, in which the temperature was kept nearly as low as that out of doors, and in which an abundance of fresh air was provided. The better condition of the fowls which immediately resulted showed the step to be taken in the right direction. Excellent ventilation is provided by this construction. Part of the front wall, or the wall on the south side, is left open or covered with muslin or good stout cheesecloth. The common custom is to use 1 square foot of cloth and 1 square foot of glass to each 18 or 20 square feet of floor space in a house 10 feet wide. Some poultry men are using cloth altogether to the exclusion of glass for all the windows, but a combination of cloth and glass is preferable.

In 1907 the Maryland Agricultural Experiment Station initiated some experiments to determine the economic value of different types of poultry-house construction. Three distinct types, with several minor variations, were used, the close, tight house, the glass-front house, and the cloth-front house. Two years of experiment indicate very strongly that the last type is the best. In the tight house the air was damp, foul, and lifeless, the plumage of the fowls become dull and rough, and the general condition was very poor. The glass-front house gave a little better results, but in the cloth-front house dampness, gases, and odors were entirely absent and the fowls were in excellent condition. The cost of construction per fowl was 34 per cent higher with the tight house than with the cloth-front. During the second year, the fowls in the cloth-front house gave a profit of 23 cents per fowl more than those in the tight house. The cloth-front house gave better results also in increased egg production, in better vitality in developing the embryo, and in producing healthier chickens.

In supplying fresh air to the fowls the danger of drafts must not be overlooked. Fowls are especially susceptible to drafts, and a little current of air blowing through the poultry house may cause the whole flock to become sick. The occurrence of drafts may generally be prevented by placing all the openings on the south side of the building, and also by placing the cloth curtains high enough above the floor so the air will circulate above the birds.

Sunlight. — Too much sunlight can never be provided in a poultry house, for the more sunlight there is, the better will be the constitution of the fowls, the more eggs they will lay, and the greater will be the financial returns. The provision of sunlight can best be accomplished by the means noted above, that of having the windows on the south side of the house. Sunlight is an excellent germicide and disinfectant, and plenty of it should keep the house fresh and sweet.

Convenience. — A number of things can be incorporated in a poultry house which will make it a convenient one. The furnishings for the fowls must include roosts and nests, and should include feed boxes, watering cans, and, for winter, a covered dust bath.

A great mistake is often made in building roosts, by placing them one above the other on inclined supports, or "horses." The domestic fowl has inherited from his wild ancestor the instinct of self-preservation, and one evidence of this is the tendency to roost as high as possible. With inclined roost supports, the fowls will seek the highest perches, and will crowd each other to such an extent as to suffocate some of them.

Roosts properly built should be horizontal, each one no higher than the rest. They should not be very high from the floor, for heavy hens are sometimes injured by falling when attempting to reach too high a perch. The height of roost should not be over 2 feet for heavy fowls, and not over 4 feet for the light, active breeds. Roosts may well be placed along the north wall of the house, and arrangement made to raise them up out of the way when cleaning. The size and shape of the perches are not unimportant, for a little care in their construction will add much to the comfort of the fowls. Two by four stock sawed in two, and with the upper corners rounded off, provides a perch amply large and strong enough and shaped so as to be most easily grasped by the fowl's claws. Figure 86 gives a cross section of a properly designed perch. The perches need not be more than 14 inches apart, but should be supported by 2 × 4 crossbars every 3 feet. The droppings can be caught just below the roosts on boards which should be so arranged as to be easily removable for cleaning purposes.

FIG 86 — Cross section of roost.

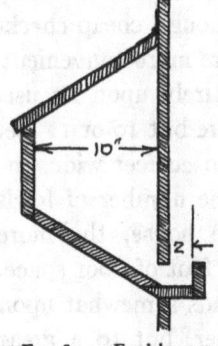

FIG. 87 — Feed box

Feed boxes may be built in below the windows along the wall of the house, and can be arranged so as to permit of filling from the outside, as shown in the diagram, Figure 87. Watering cans of a sanitary type can be purchased at any hardware store; these are portable and can be located anywhere in the house.

A dust bath is an especially desirable feature in a poultry house, for the presence of it is one of the best means of disposing of the lice difficulty. To operate to the best advantage, it should be inclosed, have a window all its own, and should be accessible by just a single small opening; this will eliminate the difficulty caused by the dust coming from an open bath. Fine road dust, or fine sifted ashes are very good materials for a dust bath, and the addition of a little lime, of tobacco dust, and of good lice powder tends to make it more effective.

General Construction. — Poultry houses are usually of light construction, scarcely ever any conditions arising which might require great strength. It is possible and entirely practical to construct poultry houses with no lumber heavier than 2 × 4's, and indeed almost all of them are constructed in such a manner. Sills, studs, plates, rafters, and necessary braces may all be of 2 × 4, and amply strong, since the house is low and not subjected to severe racking by the winds. For wall covering shiplap or drop siding is excellent. The roof may be covered either with shingles or prepared roofing, the latter being especially desirable on the low-pitch roofs entering so often into poultry-house construction. For windows ordinary barn sash can be used, if so desired, though cheap checkrail windows cost but little more and are more convenient.

The width of the house depends entirely upon its use. Many general-purpose poultry houses are but 10 or 12 feet wide; laying houses are generally 14 to 20 feet wide, the length being governed, of course, by the number of fowls to be accommodated. The wider the house, the more economically can it be built per square foot of floor space.

The most suitable style of roof depends somewhat upon the methods used by the poultry raiser, but to a great extent upon the type of house. The commonest form is the shed roof, with only one slope, to the north. This form of roof has certain advantages, inasmuch as all the water runs off at the rear and it will not absorb so much heat from the sun during the summer. The single-pitch roof should be used only where the span of the roof is less than 14 feet; otherwise sagging will result. Besides, the front would have to be unnecessarily high in order to give the roof sufficient pitch in wide spans.

The gable or A-shaped roof is also a common type, and is especially suitable in case the fowls are to be yarded in an orchard, the peaked roof admitting the drawing along of the house with the minimum disturbance of the branches of the fruit trees.

It affords a steeper pitch, which is desirable for shingle roofs, and can be used for wider spans than can the single-slope roof.

A combination of the two forms of

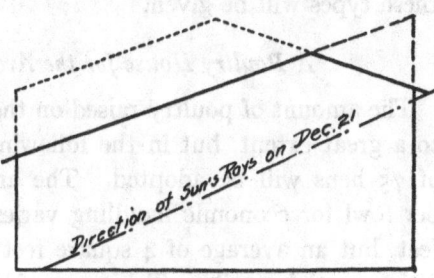

FIG. 88. — Forms of roofs.

roofs noted above has come into vogue, and seems to prove quite satisfactory. It has the advantage of both forms in that it can be used on a wide span and affords a steep pitch with less cost of siding. It is shown in Figure 88 with heavy lines, and a study of the figure will show the saving of lumber over both the other types. It will be seen that this type of roof admits the sun's rays on December 21 almost to the extreme rear of the house.

Types of Poultry Houses

Three general types of poultry houses are recognized, the classification being based upon the extent and importance of the poultry-raising business under varying conditions. On the average farm, poultry raising is not the chief business; it is a sort of side line, mainly maintained as a source of food supply. The average farmer, then, will need a permanent, general-purpose house whose cost is not excessive. On the other hand, the man conducting poultry farming as his main business must have a house large enough to produce poultry in large quantity economically; he will usually have a commercial laying house in which to winter his laying hens, and portable colony houses for brooding early chicks. Descriptions of each of these types will be given.

A Poultry House for the Average Farm

The amount of poultry raised on the average farm varies to a great extent, but in the following description a basis of 75 hens will be adopted. The amount of floor space per fowl for economic handling varies from 3 to 5 square feet, but an average of 4 square feet will meet most conditions satisfactorily. Taking a width of 14 feet as one that will adapt itself well to this type of house, the length will approximate 24 feet. The height of the building may be whatever the owner desires, but if a 10-foot 2 × 4 is used for a stud, it can be cut into two pieces, one 4 feet, the other 6 feet long, the first being used at the rear, and the latter at the front. Figure 89 illustrates the framing of this house, and gives the length of the members; all framing is of 2 × 4 stock. A modified gable roof is shown, with a low pitch, in consequence of which prepared roofing is used as a covering; a single-pitch or shed roof can be used to equally good advantage. The front view of the building shows the arrangement of the windows both for light and for ventilation. Double sash are used, the upper half being covered with muslin, the lower half being of glass. The ventilation can be admirably controlled by raising or lowering the muslin frames to suit the weather,

CONSTRUCTION OF FARM BUILDINGS

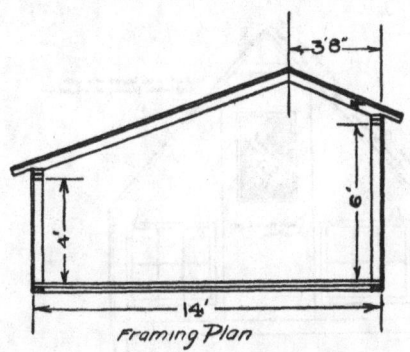
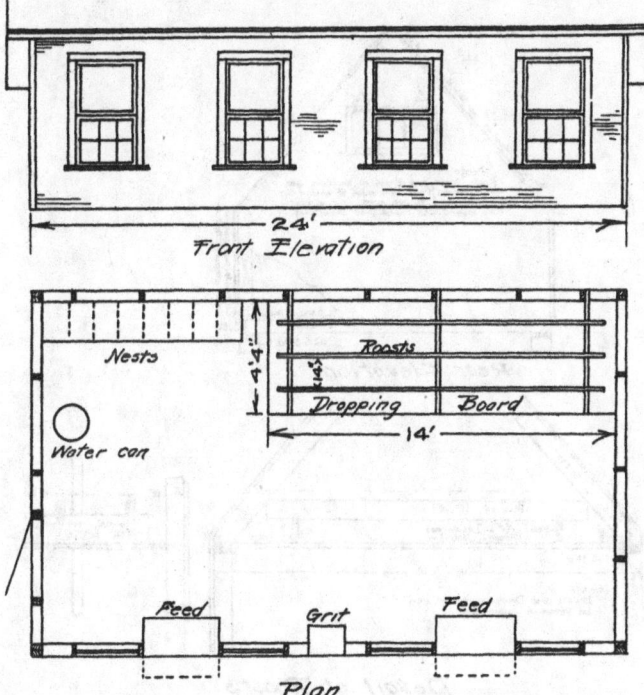

FIG 89. — A poultry house for the average farm.

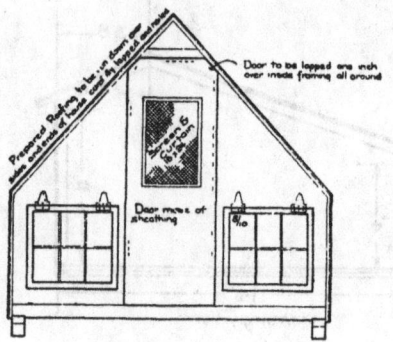
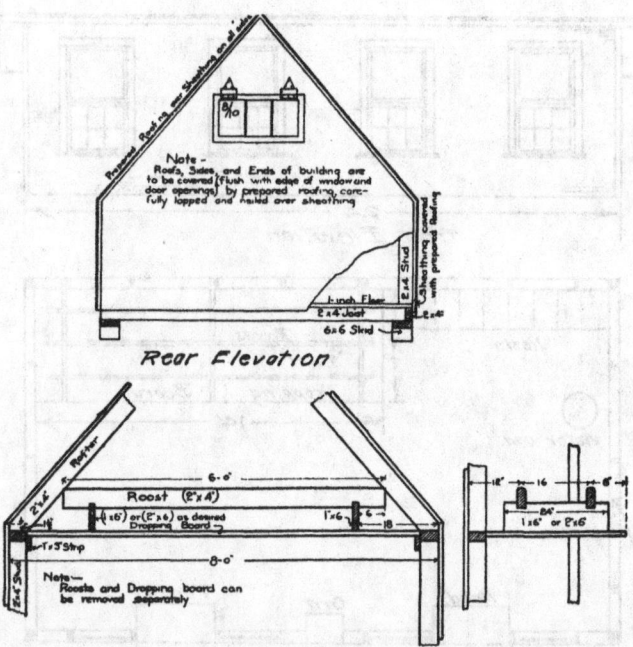

Fig. 90 a. — Portable colony house. (Iowa Agr Exp. Sta.)

CONSTRUCTION OF FARM BUILDINGS

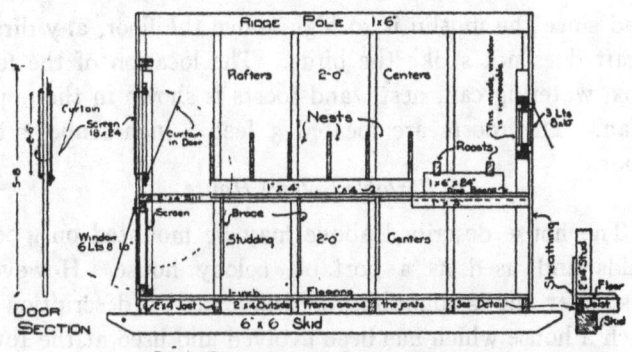

Side Elevation of Framing

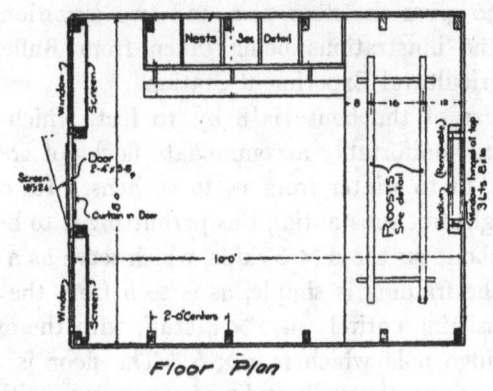

Floor Plan

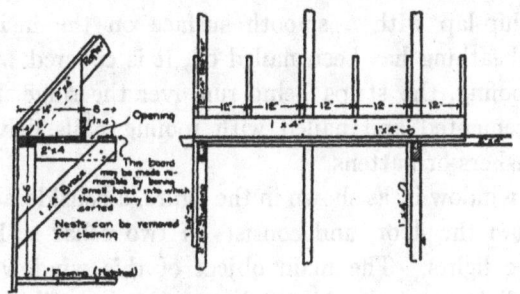

Detail of Nests.

FIG. 90 b. — Portable colony house. (Iowa Agr. Exp. Sta.)

and since the muslin is so high above the floor, any direct draft does not strike the birds. The location of the feed box, watering can, nests, and roosts is shown in the upper plan. The roosts are located 3 feet 6 inches above the floor.

Portable Colony House

The house described above may be mounted on 4 × 6 skids and used as a portable colony house. However, a smaller one is sometimes desirable, and a description of such a house which has been evolved and used at the Iowa Agricultural Experiment Station is herewith included. Figure 90 gives the floor plan and two elevations of the house, the illustrations being taken from Bulletin 132, Iowa Agricultural Experiment Station.

The size of the house is 8 by 10 feet, which is large enough to comfortably accommodate flocks of 200 to 300 chickens, or to winter from 15 to 20 hens. Its construction is light but substantial, this permitting it to be readily moved about on the 6 × 6 skids, which serve as a foundation. The framing is simple, as is seen from the illustration, consisting entirely of 2 × 4 stuff, with the exception of the ridge pole which is 1 × 6. The floor is of plain 6-inch flooring; the walls and roof are covered with a good grade of ship-lap with a smooth surface on the inside. After the sheathing has been nailed on, it is covered with prepared roofing, the strips being run over the ridge, the laps well cemented and nailed with roofing nails driven through washers or battens.

The rear window is, as shown in the figure, a considerable distance from the floor, and consists of two cellar sashes with 9 × 12 lights. The main object of this window is to provide light, and should not be used for ventilation, because the fowls roost near it and all possible drafts must be guarded against. The front windows, of which there are two, are located one on each side of the door and consist of 6-light barn sash protected on the inside by screen, and fitted with hinges at the top so they may be swung up and out. The door is provided with a screened opening at the top with a ventilating curtain fitted with a hinged frame on the inside.

The construction and location of the roost, nests, and dropping board are shown in the detail drawing accompanying. Both may be so installed as to be removable, this making cleaning easier.

In moving the building about, care should be taken not to subject it to any severe or undue racking. If the two skids are connected to each other by a chain, and a horse is hitched to the chain, the strain will tend to twist the whole structure. To avoid this it is a good plan to hitch a horse to each skid, or fasten a stiff spreader between the skids.

The estimated cost of such a house is approximately $40.

A Commercial Laying House

For commercial poultry raising a larger and more elaborate equipment in the way of buildings is necessary. The drawings of such a house are shown in Figure 91; while these drawings illustrate only a single section 18 feet square, the house may be made as large as desired simply by duplicating the sections, and making a long building.

The roof of this structure is of the modified gable type, with a pitch steep enough to admit the employment of shingles as a roof covering. The rear studs are 4 feet 6 inches in height, while those in front are 7 feet 6 inches, each pair of studs utilizing the whole of a 12-foot 2 × 4.

The double 2 × 4 used as a plate brings the height of the braces up to 7 feet 10 inches, a good working height. The walls are covered with a good grade of ship-lap or drop siding whose edges are painted with white lead and made close and tight when nailed on. The floor should be made of cement with an asphalt coating as hereinbefore described, and a thick layer of straw is applied upon it to eliminate

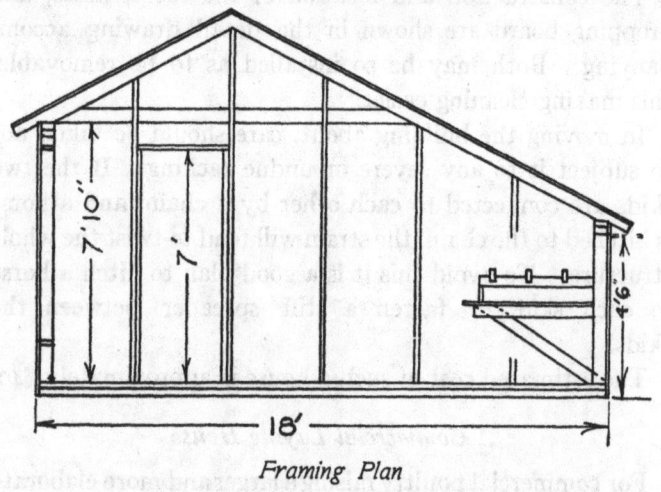

Framing Plan

Fig. 91 a. — Commercial poultry house.

any tendency toward cold. All furnishings are portable, and nests and boxes have sloping tops, which precludes the possibility of their being used for roosts and becoming foul from droppings.

The structure is ventilated by means of the muslin curtains in the upper sash of the four double-hung windows in front. The door is made with a long single sash, which admits an abundance of light into the house in addition to that coming through the windows.

CONSTRUCTION OF FARM BUILDINGS

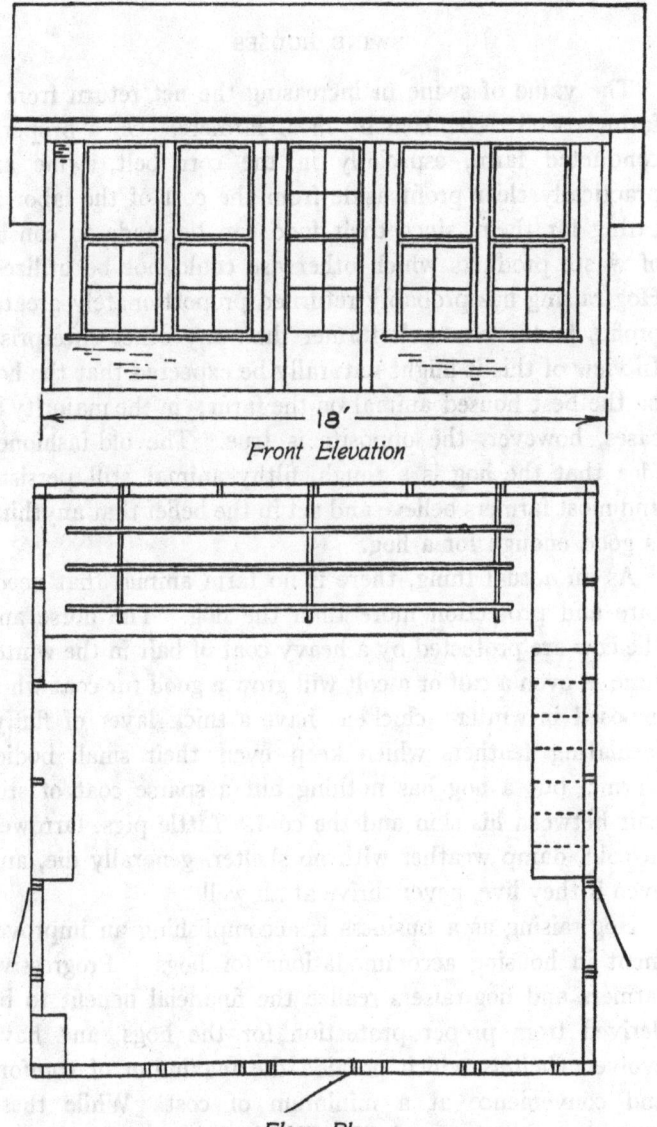

FIG 91 b — Commercial poultry house.

SWINE HOUSES

The value of swine in increasing the net return from a farm is well recognized by most farmers. On a properly conducted farm, especially in the corn belt, swine are practically clear profit aside from the cost of the labor in caring for them, since their food can be made to consist of waste products which otherwise could not be utilized. Hog raising has probably returned proportionately greater profits to the corn-belt farmer than any other enterprise. In view of this it might naturally be expected that the hog be the best housed animal on the farm; in the majority of cases, however, the opposite is true. The old-fashioned idea that the hog is a tough, filthy animal still persists, and most farmers believe and act in the belief that anything is good enough for a hog.

As an actual thing, there is no farm animal that needs care and protection more than the hog. The horse and the cow are protected by a heavy coat of hair in the winter time — even a calf or a colt will grow a good fur coat when exposed in winter; chickens have a thick layer of fluffy, insulating feathers which keep even their small bodies warm; but a hog has nothing but a sparse coat of stiff hair between his skin and the cold. Little pigs, farrowed in cold, damp weather with no shelter, generally die, and even if they live, never thrive at all well.

Hog raising as a business is accomplishing an improvement in housing accommodations for hogs. Progressive farmers and hog raisers realize the financial benefit to be derived from proper protection for the hogs, and have evolved shelters which provide the maximum of comfort and convenience at a minimum of cost. While these structures are to be found principally on farms where hog raising is the chief business, they can easily be so modified or reduced in size as to meet with the approval of the farmer conducting a general-purpose farm where hogs are but one source of financial returns.

The essentials of a swine house are comfort for the animals under all conditions, convenience for the caretaker in feeding and handling them, and good sanitation. The house must provide sufficient warmth in cold weather to keep the swine in good condition, and must provide shade on the hottest days of the summer. It must be so arranged as to permit of feeding the hogs easily and expeditiously, and of handling them quickly and with the least amount of disturbance. Sanitation is especially to be emphasized in swine-house construction, and some attention paid to this particular will be amply rewarded later on in healthier swine and consequently greater financial returns.

Swine houses must be built to include such arrangements as will initiate and maintain a tendency to eliminate disease instead of fostering or developing it as is so often the case. Since most of the diseases which seriously affect swine are germ diseases, it follows that any construction which will prevent the ingress and development of germs is the most advantageous form to follow. Sunlight is one of the commonest, most effective, and withal, the cheapest, germicide known; the swine house, then, should be built with the provision made for a maximum amount of sunlight at the times when it is the most needed.

Hog raisers usually desire that two litters of pigs be raised each year, and it has been found that the times of farrowing of these litters, for the best results, should be about March 1 and August 1. The pigs farrowed at the latter time will generally thrive well, weather conditions being in their favor; but the spring litter is often seriously handicapped by the cold weather which is very likely to occur at that time. To offset this disadvantage, every bit of sunlight must be utilized, and the windows should be so arranged as to height and location that this can be accomplished. With an abundance of sunlight, and with walls and floors as nearly vermin proof as possible, sanitation is taken care of, as far as the building is concerned.

Swine generally require a great deal of care and special attention. Beginning at farrowing time, the sow must be isolated; this requires that individual pens must be provided for the pregnant sows, and kept for them until the pigs are a week or ten days old. Brood sows with their litters should be given a small yard of their own in which they can be kept until the pigs have learned to recognize their mother, when they can be turned out into the general feeding lot. In order that some degree of cleanliness in their food be maintained, feeding floors of concrete must be built. The breeder of high-grade swine usually desires to have some place in which he can advantageously exhibit his stock to a prospective purchaser. A dipping vat is an essential part of the equipment of a good hog farm. Hogs require a quantity of good, clean feed, well prepared and given to them in the most economical form.

All the things enumerated above tend to bring the unit cost per animal rather high in providing hog-raising equipment, and unless the farmer is careful, the cost may run so high as to absorb the greater part or even all of the possible profits. It is very easy to get too much expense into any farm building and the swine house is no exception; no one can afford for any purpose a building so expensive that interest and depreciation will more than counterbalance its value as a shelter. The maximum cost should never be over $40 per pen, and indeed very efficient swine houses can be built for $25 to $30 per pen, at present prices of building materials.

Most of the construction details of swine houses have been almost standardized, which is not true of many other farm buildings. Practically all hog breeders are agreed that a pen 6 × 8 feet is amply large for a sow and her litter; indeed, 5 × 8 feet is a common size. The partitions between pens should be so contrived as to permit of throwing the whole house or any part of it into a large pen. The best floor for a swine house is perhaps the earth floor, but this is very hard to keep in a sanitary condition; wood and concrete floors have been used, but each has its disadvantages, the wood being shortlived, affording a harbor for disease germs, and the concrete being too cold in the spring at farrowing time. A solution of the difficulty is rather hard to find. Where the winters are at all severe, it is desirable that the walls be made double, either by putting a double layer of boards on the exterior of the studs or by boarding up both inside and out.

Types of Swine Houses

Several types of swine houses have been constructed and used with varying degrees of success, and all of them have their advocates. Two general types stand out rather prominently, however, and seem to meet with the approbation of progressive hog raisers. These types are the individual houses, and the large houses with individual pens. Modifications of these types are numerous, the modifications resulting from the needs and ideas of individuals building them.

Individual hog houses, or cots, as they are sometimes called, are built in many different ways. The commoner methods of constructing houses are illustrated in Figures

206 FARM STRUCTURES

92 and 93. The first figure illustrates a four-walled variety, collapsible, so that it may be taken down, removed, and again erected with a very small amount of labor. Figure 93 illustrates another kind, which has two sloping sides reaching from the ridge to the ground, forming a sort of tent-shaped structure; these may be constructed with the four walls and floor so arranged as to be collapsible, or they may be mounted on skids and thus made portable, a horse being required to draw them about. Some styles have a window in the front and above the door; all should have a small door in the rear and near the ridge for ventilation.

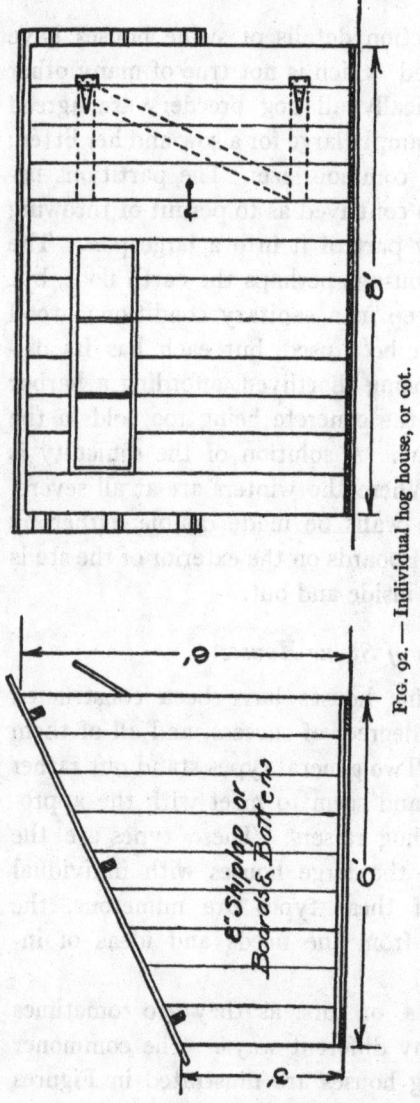

FIG. 92. — Individual hog house, or cot.

There are a number of points to be enumerated in favor of the individual type of swine house. Each sow may be isolated at farrowing time, and for some time afterward; the houses may be placed at the end of the lot farthest away from the feeding floor so the sows may be compelled to exercise; the danger of spreading disease is reduced to a minimum, and should the location of any house become unsanitary, it can easily be moved to another location.

Large houses, if properly built, have some advantageous features that commend them to careful hog raisers. Excellent sanitation can be accomplished in a substantially built house of this type, especially where a concrete floor is used. The swine may be handled very easily and conveniently, and the plan of arranging the pens may be such that feeding may be done with the minimum of labor.

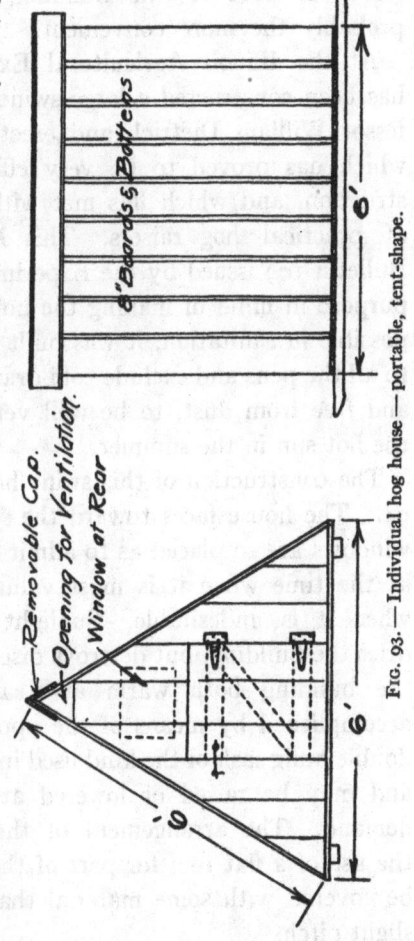

FIG. 93.—Individual hog house—portable, tent-shape.

With portable partitions, the house may be divided into farrowing pens, or the partitions may be omitted, thus providing for an abundance of light in a house of this type. Bins for storing feed may also be included, either on the same floor or in a small loft over part of the building, though the former is probably the more convenient.

At the Illinois Agricultural Experiment Station there has been constructed a large swine house planned by Professor William Dietrich and erected under his direction, which has proved to be very efficient in point of construction, and which has met with the marked approval of practical hog raisers. This has been described in Bulletin 109 issued by the Experiment Station. With the purpose in mind of making the house as nearly perfect as possible in sanitation, it was built so as to admit sunlight to all the pens and exclude cold drafts in winter, to be dry and free from dust, to be well ventilated and to exclude the hot sun in the summer.

The construction of this swine house is shown in Figure 94. The house faces toward the south, and both tiers of windows are so placed as to admit a maximum of sunlight at the time when it is most valuable, and to exclude it when it is undesirable. Sunlight not only warms and dries the building, but destroys disease germs, thus making the building both warm and sanitary. Ventilation is accomplished by means of the upper windows, which are double hung sash of the kind used in residence construction, and may be raised or lowered at will as circumstances demand. The arrangement of the windows necessitates the use of a flat roof for part of the building, which must be covered with some material that will shed water at a slight pitch.

The house is 30 feet wide, and an 8-foot alley running down the center divides it into two parts of equal width, each of which is divided into 9 pens, space being left at one

CONSTRUCTION OF FARM BUILDINGS 209

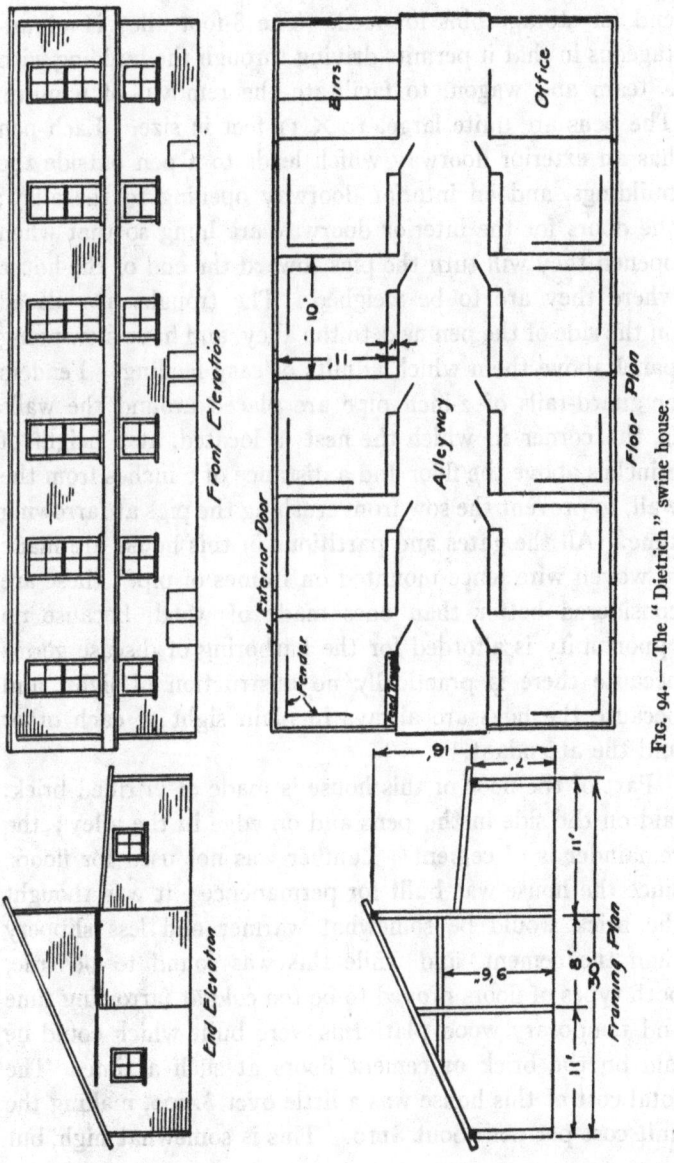

FIG. 94.—The "Dietrich" swine house.

end for storage bins for feed. The 8-foot alley is advantageous in that it permits driving through the building with a team and wagon, to facilitate the removal of manure. The pens are quite large, 10 × 11 feet in size. Each pen has an exterior doorway which leads to a pen outside the buildings, and an interior doorway opening to the alley; the doors for the interior doorway are hung so that when opened they will turn the pigs toward the end of the house where they are to be weighed. The troughs are placed on the side of the pen next to the alley, and have a swinging panel above them which admits of easy feeding. Fenders or guard-rails of 2-inch pipe are placed around the walls in the corner in which the nest is located, at a height of 8 inches above the floor and a distance of 6 inches from the wall, to prevent the sow from crushing the pigs at farrowing time. All the gates and partitions in this house are made of woven wire fence mounted on frames of pipe; these are considered better than ones made of wood, because no opportunity is afforded for the harboring of disease germs because there is practically no obstruction of light, and because the hogs are always in plain sight of each other and the attendant.

Part of the floor of this house is made of vitrified brick, laid on the side in the pens and on edge in the alley; the remainder is of cement. Lumber was not used for floors, since the house was built for permanence; it was thought the brick would be somewhat warmer and less slippery than the cement, and while this was found to be true, both types of floors proved to be too cold at farrowing time and temporary wood platforms were built which could be laid on the brick or cement floors at such a time. The total cost of this house was a little over $2000, making the unit cost per pen about $110. This is somewhat high, but the house as built on an ordinary farm would have smaller pens, and a less expensive construction generally, the cost being materially reduced thereby.

A rather unique design of a large swine house is shown in Figure 95. Mr. W. H. Smith, of the Illinois Agricultural Experiment Station, is the originator of the design, and he has used the house successfully for a number of years. It has a number of distinct advantages, among which are the following: economy in construction, as shown by an actual case in which a house with six 18-foot sides cost approximately $400; efficiency in operation, the handling and feeding of the hogs being accomplished very easily from the

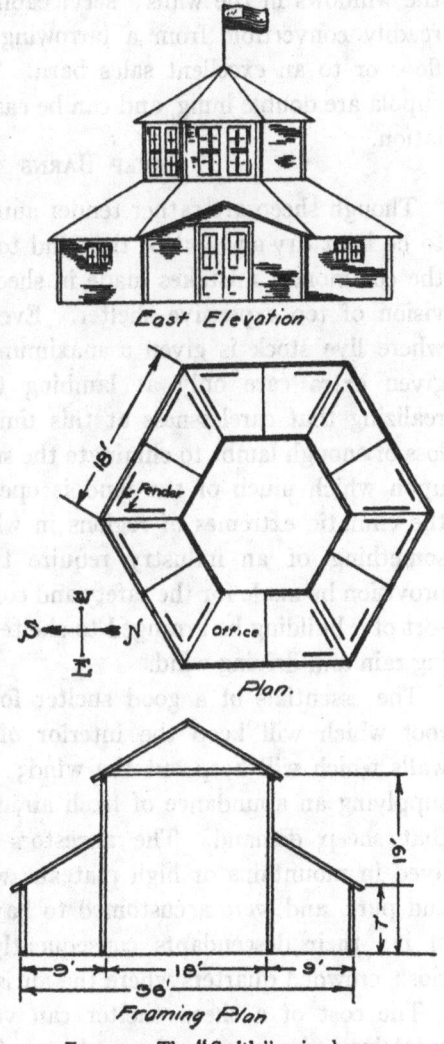

FIG. 95.— The "Smith" swine house.

central space; an abundance of light, since the sun comes in from four sides of the cupola, as well as from the windows in the walls; serviceability, the house being readily convertible from a farrowing house to a feeding floor or to an excellent sales barn. The windows of the cupola are double hung, and can be easily opened for ventilation.

SHEEP BARNS

Though sheep are rather tender animals, they need only to be kept dry and out of the wind to thrive; and one of the commonest mistakes made in sheep raising is the provision of too expensive shelter. Even in the old world, where live stock is given a maximum of care, sheep are given extra care only at lambing time, the shepherds realizing that carelessness at this time may result in the loss of enough lambs to eliminate the small margin of profit upon which much of the land is operated. In America, the climatic extremes of regions in which sheep-raising is something of an industry require that a more careful provision be made for the safety and comfort, and that some sort of a building be arranged to shelter them from drenching rain and driving wind.

The essentials of a good shelter for sheep are a tight roof which will keep the interior of the building dry; walls which will keep out the wind; and some means of supplying an abundance of fresh air, for this is one thing that sheep demand. The ancestors of sheep generally lived in mountains or high plateaus where the air is fine and pure, and were accustomed to having their lungs full of it; their descendants consequently cannot thrive in close, crowded quarters where the air is impure and bad.

The cost of a sheep shelter can very easily be made excessive. An average horse barn of good construction will cost approximately $50 per horse, or the shelter will cost about one fourth the value of the horse; on the same basis, assuming the value of a sheep to be about $7.50, the shelter for one sheep should cost $1.85, and since about 6 square feet of floor space is necessary for a single sheep, the cost per square foot of floor space for a well-built sheep shelter should not be more than 30 cents. As a matter of fact, sheep barns are often built in which the cost per square foot is twice or even three times 30 cents, but usually these barns have storage room for large quantities of feed and require heavier and more expensive construction. The question as to whether a simple shed with no storage capacity is better than a high barn in which provision is made for extensive storage, must be settled by surrounding conditions and by personal preference. The governing factors in the problem are the cost of construction of three types of structures, namely: a simple sheep shed, a simple storage house, and a combined shelter and storage house, and the cost of labor as influenced by the accessibility of the feed. While it probably will be found considerably cheaper to construct two separate simple buildings, one for shelter and the other for storage, the additional labor of transporting the feed from the storage shed to the sheep may more than counterbalance the saving resulting from the cheaper buildings.

In any building constructed for sheltering sheep, there are a number of features which are more or less essential, and one of the most distinctive of these is a separate compartment in which the pregnant ewes can be isolated at lambing time. This space is divided up by means of portable hurdles into small pens four or five feet square, each pen accommodating one ewe; when the lambing season is over the hurdles can be removed and the compartment used as a feeding room for the older lambs. It is advisable to have adjacent to the lambing room another room used as the shepherd's quarters, which is furnished with a stove; in very inclement weather the door between the lambing room and the shepherd's room can be opened, and sufficient warmth supplied to the young lambs to keep them from suffering from the cold.

Some good sheep barns have been built which were divided up into permanent pens, but under most conditions it is well to so arrange the plans as to make it possible to keep the floor clear when need arises, since the main purpose of the barn will be to shelter a large number of sheep in bad weather and at feeding times. Should it be found necessary to have some of the sheep kept separate, pens can easily be constructed by using either portable hurdles or feed racks as partitions. When the floor is made so as to be unobstructed as much as possible, it greatly facilitates feeding and cleaning. Earth floors are entirely practical in sheep barns, and when kept properly cleaned, are very satisfactory. It is common to let manure accumulate to a considerable depth before it is removed; this accumulation of manure is not usually attended by offensive odors, since the constant stirring by the small feet of the sheep tends to keep it from heating. In fine weather it is of advantage to have the stock out of doors, and at times such as this and in summer time it is desirable to have an outdoor paved feeding yard.

The ventilation of the sheep barn can best be accomplished by means of doors and windows so constructed as to admit of flexible control, though the King system can often be used to advantage. In some large and successful sheep sheds the exterior swinging doors are divided into two parts, each mounted on separate sets of hinges, and in fine weather, even in rather cold weather, the upper half of the doors on all sides is left open; should there be a cold, damp wind blowing from some direction, the doors that are on that side may be closed, leaving the ones on the leeward side open; while in very inclement weather, which would last for a comparatively short time, all the doors may be completely closed. The same efficiency can be obtained by using windows as ventilators, the windows consisting of barn sash mounted on hinges at the bottom; any degree of ventilation can be secured by this means, and it has the advantage of deflecting the air currents upward, thus avoiding drafts to a great extent.

In feed racks, as in everything else, there is a right and a wrong method of construction.

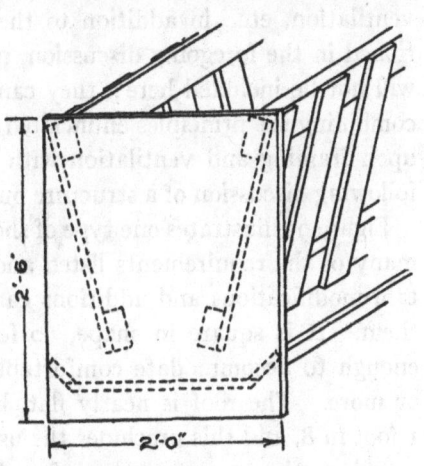

FIG. 96. — Feed rack for sheep.

In Figure 96 is illustrated a type of rack perfected by Professor W. C. Coffey of the Illinois Experiment Station, which fulfills the requirements of a successful rack. It is large enough to admit of placing a considerable quantity of feed within it, the slats are spaced far enough apart so the sheep can get its head in and eat without having to tear the feed out upon the floor, and it has a close tray below for the fine feed which will catch any wastes from the rack above. It can be made either single or double, so stock can feed from one or both sides:

the length can be made such that it will exactly fit in between adjacent interior posts, thus serving as an excellent partition. Ample watering facilities should be provided the sheep, and troughs should be located at various convenient points, and kept filled with water clean enough for human use.

Since the design of a combined sheep shelter and storage barn embodies several features, such as special framing, ventilation, etc., in addition to the distinctive ones mentioned in the foregoing discussion, plans of such structures will not be included here; they can be arranged easily by combining the principles enunciated in subsequent chapters upon framing and ventilation with the ideas given in the following discussion of a structure built for shelter only.

Figure 97 illustrates one type of sheep shed which satisfied many of the requirements listed above, and which by certain modifications and additions can be made to fit all of them. It is square in shape, 60 feet on each side, large enough to accommodate comfortably a flock of 500 sheep or more. The roof is nearly flat, having a slope of only 1 foot in 8, and this precludes the use of shingles for a roof covering, tin or prepared roofing being necessary. The framing consists of three pieces of 2×6 stuff nailed together to form posts, supported at the bottom by a concrete foundation along the exterior walls and by concrete piers in the interior; the posts are spaced 12 feet apart in both directions. Across the tops of the posts are laid three 2×6's, two of them vertical and one horizontal, upon which are placed the 2×6 rafters. The walls are made of horizontal drop siding, this necessitating the use of studding between the exterior main posts. The building is well lighted and ventilated by means of the windows on all sides and in the raised interior bent at the top. The

CONSTRUCTION OF FARM BUILDINGS 217

portable feed racks are of such a length that they will fit exactly between the framing posts and thus serve admirably as partitions when there is need of pens. The large pens formed by the feed racks can be further subdivided

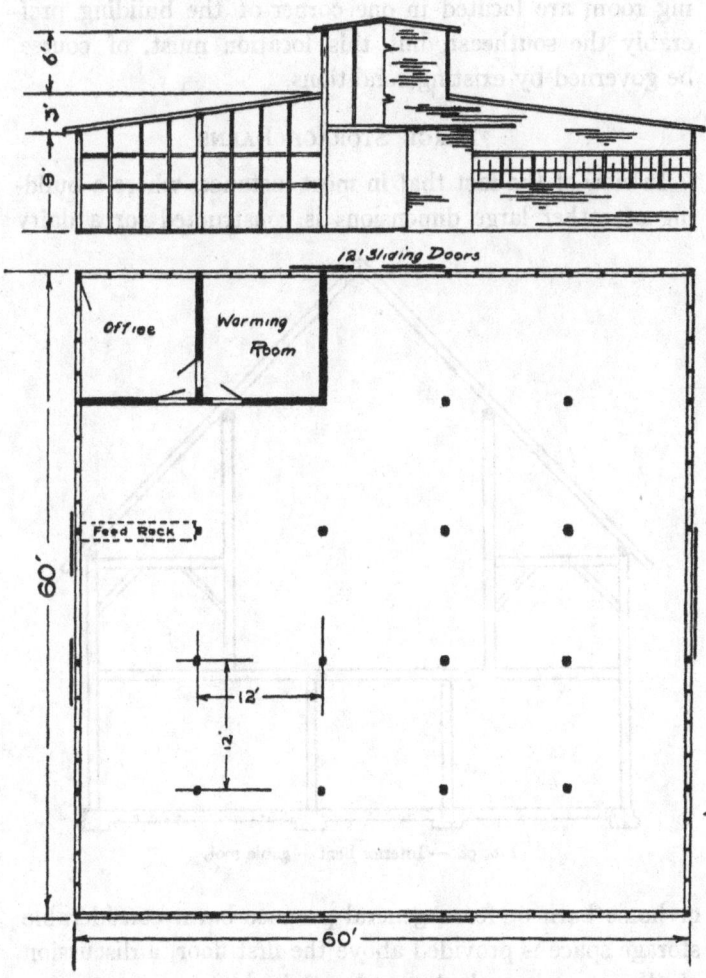

FIG. 97. — Sheep shed.

218   FARM STRUCTURES

into smaller pens by portable hurdles. The exterior doors are made high enough to admit a team and wagon, and a clear driveway can be maintained through the entire shed in both directions. The shepherd's quarters and the warming room are located in one corner of the building, preferably the southeast, but this location must, of course, be governed by existing conditions.

Large Storage Barns

In view of the fact that in most instances where a building of rather large dimensions is constructed for a dairy

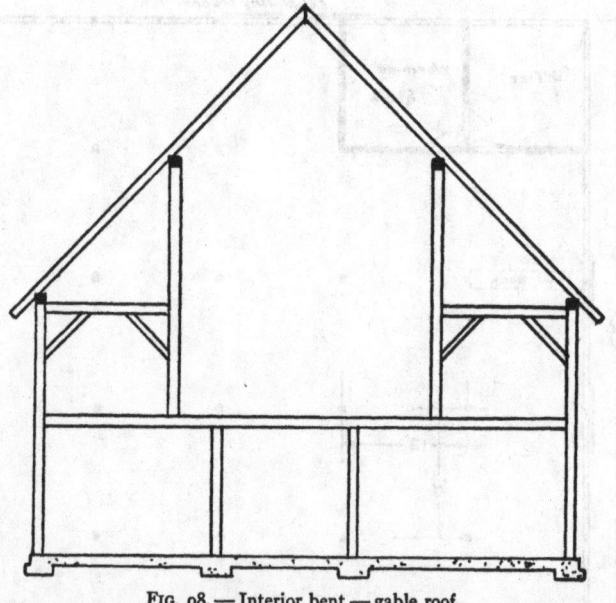

Fig. 98. — Interior bent — gable roof.

or horse barn or for a general purpose barn, considerable storage space is provided above the first floor, a discussion of the general methods employed in such a construction

CONSTRUCTION OF FARM BUILDINGS 219

will be taken up. The principles and construction methods given in this discussion will apply equally well to any of the above-mentioned barns, while the details which are applicable to only one specific type of barn will be considered in subsequent discussions.

Framing. — There are two entirely different systems of framing employed in the construction of ordinary large

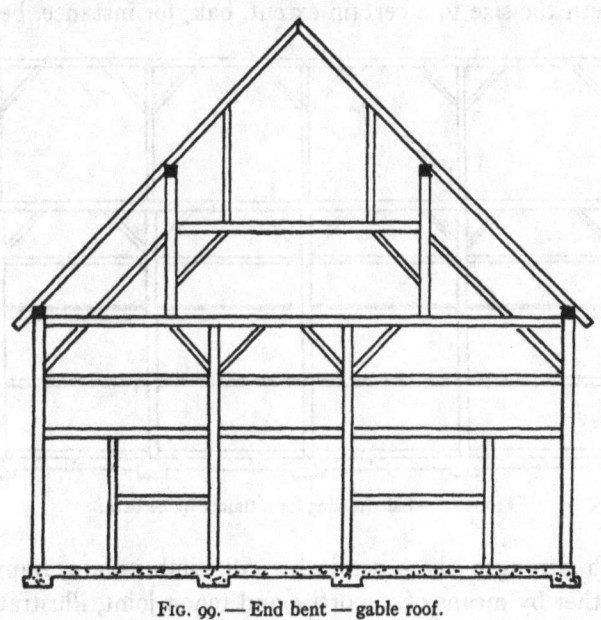

FIG. 99. — End bent — gable roof.

barns, the timber frame and the plank frame. The former was used almost exclusively in the earlier days when timber was cheap and could be obtained in almost any desired size or length; the latter type has been developed to reduce the cost of construction, and accomplishes this by using lumber which is only 2 inches in thickness, and which of course can be obtained at a much lower cost than that of large timbers.

Timber Framing. — Figures 98, 99, 100, and 101 illustrate the details of the timber frame. As is seen, the frame consists of large rectangular timbers of varying sizes, from a 12 × 12 sill at the bottom, to 4 × 4 braces, and 2 × 4 or 2 × 6 rafters. The sizes of the timbers will, of course, vary with the size of the barn and with the load to which they are to be subjected; the kind of timber used will also govern the size to a certain extent, oak, for instance, being

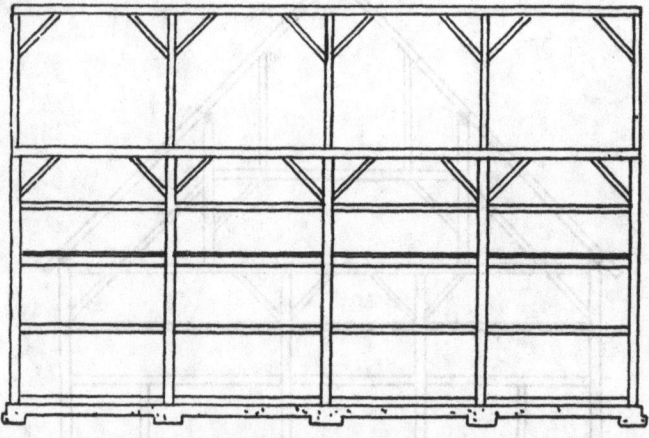

FIG. 100. — Side framing for a timber frame barn.

much stronger than hemlock. All timbers are framed together by means of a mortise and tenon joint, illustrated in Figure 102, through which a wooden dowel pin is driven. These pins are given a long taper and the holes in the tenon and those through the mortise are given $\frac{1}{8}$-inch "draw" in such direction as will tend to pull the shoulder of the timber on which the tenon is formed close up against the timber in which the mortise is cut; that is, the distance from the joint to the dowel hole is $\frac{1}{8}$-inch greater in the mortise than in the tenon.

The large timbers used for crossbeams and ties should be whole sticks, not spliced, but the sills and plates may be spliced at every bent. (The term "bent" is used to cover one of the units of framing extending across the building; it is also sometimes taken to mean the space included between the framing units.) Sometimes long

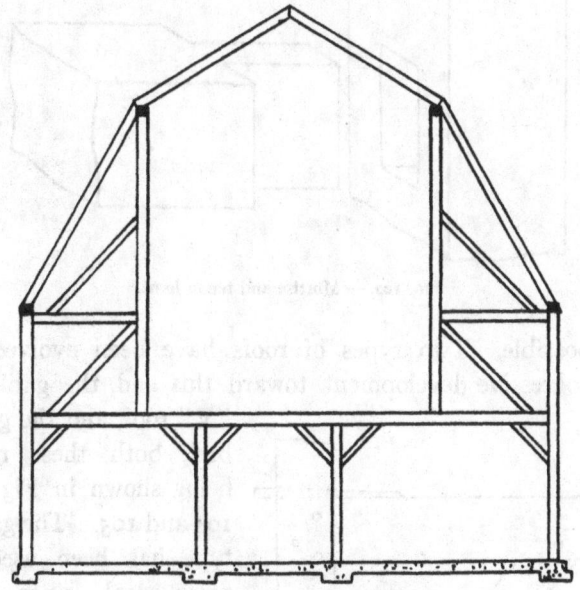

FIG. 101. — Interior bent — gambrel roof.

timbers are very difficult to obtain, in which case the width of the barn can be adapted to the length of the timbers obtainable, making one strong splice at the center, if necessary. In each of the angles formed at the intersection of two large timbers should be placed a diagonal brace 3 or 4 feet in length, the ends of which are connected with the main timbers by means of a mortise joint held with a dowel pin, as shown in Figure 103; this figure also shows a very efficient method of joining a crosstie and a post.

One point especially worthy of note for storage barns is the effort to include as much under a given amount of roof

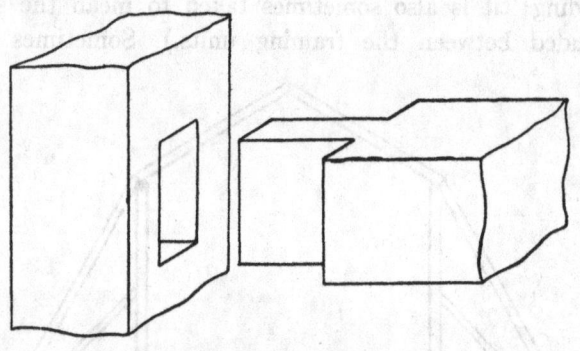

FIG. 102. — Mortise and tenon joint.

as possible. Two types of roofs have been evolved in a progressive development toward this end, the gable or

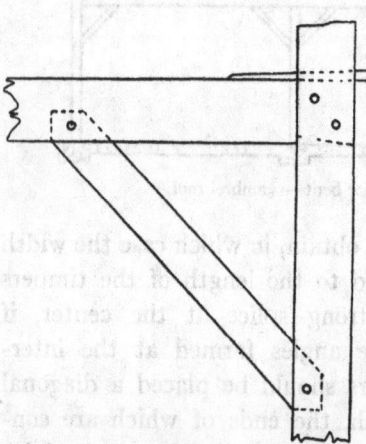

FIG. 103. — Mortised joint in brace and crosstie.

"V" roof, and the gambrel, both these roofs being shown in Figures 104 and 105. The gable type has been used a great deal, since the framing of the bents for this type of roof is quite simple; however, in attempting to employ the principle enunciated above, it was soon found that the gable roof did not give greatest amount of room under a given amount of roof, and in consequence the advantages of the gambrel roof came into wider recognition. A comparison of the two types will make the difference between them manifest; the mow is only half filled when filled even with the plate, and the gambrel roof provides a much larger space above the plates than does the gable roof, even though it have a steep pitch. The lower pitch of the

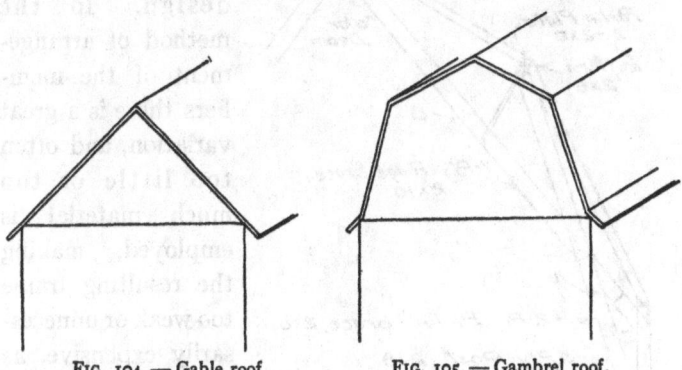

FIG. 104. — Gable roof.   FIG. 105. — Gambrel roof.

gambrel roof is so nearly vertical that it is in effect almost a wall.

The timber frame makes a very strong and substantial structure, and when the sills are kept off the ground so that they do not rot and when the superstructure is properly protected by the exterior wall covering, such a frame retains its strength for many years. It is similar to the braced frame used in residence construction, examples of which can be found that are more than a century old. The chief disadvantages of the timber frame lie in the high cost of large sticks of timber, in the difficulty of handling them, and in the fact that many of the timbers have their tensile and shearing strength reduced by 50 per cent or more by the reduction in cross-section area necessitated in the mak-

224   FARM STRUCTURES

ing of a mortise and tenon joint. The controlling strain, however, is usually a transverse one, rather than one of tension or shear.

Plank Framing. — The plank frame, since it is a comparatively modern development, has perhaps as yet not reached its most perfect and economical design. In the method of arrangement of the members there is a great variation, and often too little or too much material is employed, making the resulting frame too weak or unnecessarily expensive, as the case may be. Because of the special bracing necessary to give the plank frame sufficient strength, the gambrel roof adapts itself peculiarly well to this type of framing, and is used almost exclusively. A gambrel roof supported by plank framing properly designed can be entirely supported by the exterior wall posts alone, no interior posts being necessary.

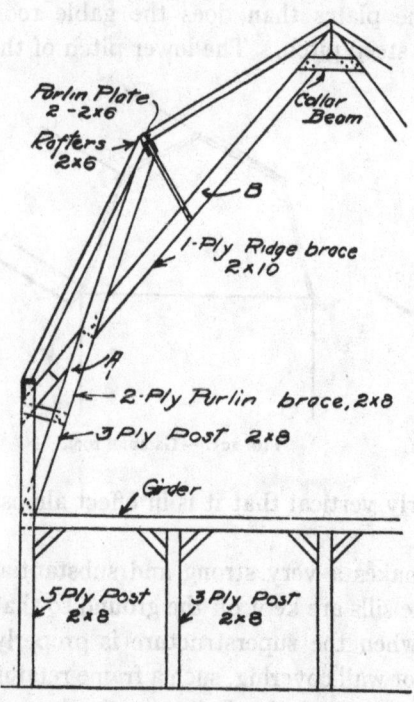

FIG 106. — Plank framing of interior bent.

The plank frame usually consists of a series of units, or bents, not more than 12 feet apart, each unit comprising a vertical post at each side and the braces, struts, etc., necessary to construct a sort of a cantilever truss; these separate bents are unified and bound together by plates nailed to the tops of the posts, by purlin plates at the break in the roof, and by subsidiary members, such as nailing girts, braces, etc. The end bents are usually framed in an essentially different manner from that of the interior bents. The interior bents are so constructed that the space above the second floor is practically unobstructed, a very desirable and more or less necessary feature, but this necessity is obviated in the end bent. Since the end wall of a barn is a large vertical one without supports or braces to resist any lateral pressure such as that of a high wind, the sticks used in framing the end bent must be arranged to give the wall the greatest possible strength and rigidity.

FIG. 107. — Location of posts.

In Figure 106 is illustrated one arrangement of the members of an interior bent of plank framing which has been found to be especially strong and practical. The description of such an arrangement for barns varying in width from 30 to 36 feet follows.

The side wall posts are built up of three 2 × 8 pieces spaced two inches apart, and extending from the floor to the plate; in the open spaces are placed 2 × 8 pieces, of a length equal to the desired clearance between floor and ceiling, usually 8 feet, and the whole thoroughly fastened together with spikes, the free use of which throughout the framing is desirable. Thus the post is made a solid 8×8 for the first 8 feet of its length, thoroughly substantial.

On the top of the 8-foot pieces are placed the girders, which support the joists. The girders themselves are supported at intervals not exceeding 12 feet by interior posts built up of three 2×8's equal in height to 8 feet plus the width of the girder, and spaced 2 inches apart, so the members of the girder may fit in between the members of the post as shown in Figure 108; or the height may be 7 feet 10 inches, in which case a flat 2-inch block covers the end of the posts and the girders are placed on this. Both the girders and the joists are designed by using the following formula for determining the size of beams subjected to a transverse strain:

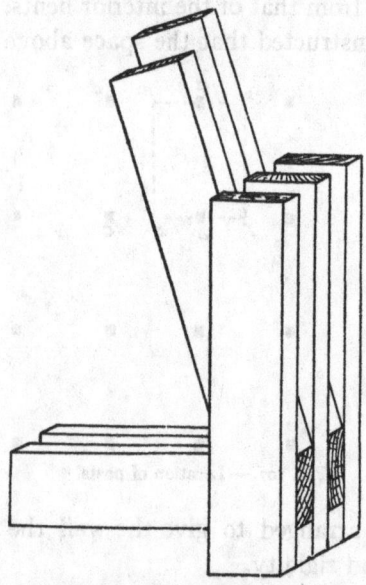

FIG. 108. — Detail of post and girder in plank framing.

$$L = \frac{2bd^2A}{S}$$

where L = safe load in pounds,
b and d = total breadth and depth respectively *in inches*,
A = 100 for oak or hard pine, 60 for soft pine,
and S = span or length of beam, *in feet*.

To illustrate the use of this formula, the size of the joists and girders of a certain barn are calculated as follows:

Let Figure 107 represent the location of the posts on the interior of the first floor. The distance apart of the bents may vary from 8 to 12 feet, and of the posts in the bent from 6 to 14 feet. Of course, the depth of the girders as well as of the joists should be the same in all parts of the barn, to maintain the level of the mow floor, consequently, we should make the design for that part supporting the greatest weight. Supposing the mow to be filled with hay, the 12-foot girder extending from a to b will support the weight of all the hay in the rectangle $efgh$; this weight may be 6000 pounds, depending upon the compactness of the hay. The formula will then be, using 2 × 10 hard pine planks for the girder:

$$6000 = \frac{2 \times b \times 100 \times 100}{12}$$
$$b = 3.6 \text{ inches}$$

or the approximate equivalent of two 2-inch widths of plank. The girder, then, will be composed of two pieces of hard pine, 2 × 10 in size.

The 10-foot joists in the rectangle $abcd$ will support the same load, and since they are to be spaced 2 feet apart, it is evident that there will be six of them in this rectangle with a total breadth of 12 inches. Substituting the known values in the formula:

$$6000 = \frac{2 \times 12 \times d^2 \times 100}{10}$$
$$d^2 = 25, \text{ and}$$
$$d = 5.$$

Since a plank 5 inches wide is usually not obtainable, the next larger standard size plank is used, a 2 × 6. This size joist will be used throughout the entire floor, for though it may be unnecessarily strong over certain spans, the spacing of the joists must not exceed 2 feet on account of the liability of the floor boards to break should they be supported at wider intervals.

The plate is constructed of two pieces of 2 × 8 stock laid flat on the top of the exterior post, or better, with one piece laid flat and with the other set in vertically beneath the flat plank in order to give the whole plate greater stiffness. This arrangement is shown in the illustration of the method of joining members in the plate.

Referring to Figure 106, the framing of the roof truss proper is seen to consist of two principal braces, *a*, the *purlin brace*, and *b*, the *ridge brace*. The purlin brace, for barns 36 feet or less in width, is composed of two pieces of 2 × 8 spaced 2 inches, the lower ends of which are inserted into the spaces between the members of the post, and rest on the girders; the upper ends are notched to receive the *purlin plate*, shown in Figure 109, which consists of two pieces of 2 × 8, which are laid on edge to give the greatest possible rigidity, and which may be spaced two inches to admit of a short diagonal brace extending down to the ridge brace. The ridge brace

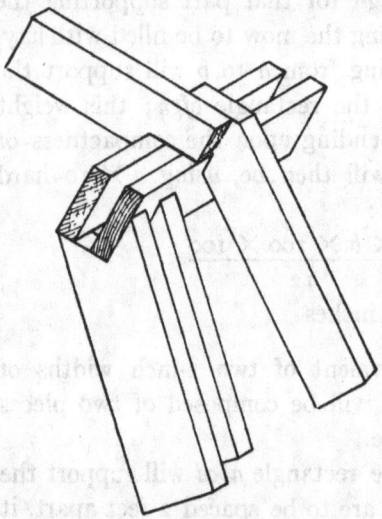

FIG. 109 — Framing at purlin plate.

CONSTRUCTION OF FARM BUILDINGS 229

itself consists of a single piece of 2 × 10 for short-span barns; it meets the corresponding brace from the other half of the truss at the ridge, while at its lower end it may be notched over the plate or brought below the plate. In the latter event, since it comes between the two members of the purlin brace, it will strike against the center member of the exterior post and will have to be bent slightly in order to enter one of the spaces; perhaps a better arrangement is effected if it is cut to fit closely against the center member, and a short strip nailed on each side, projecting beyond the end of the ridge brace and extending into the spaces of the vertical post. A short strut or two extending from the purlin brace to the ridge brace and to the post will aid materially in stiffening the truss. Figure 110 shows the method of joining the various members at the plate.

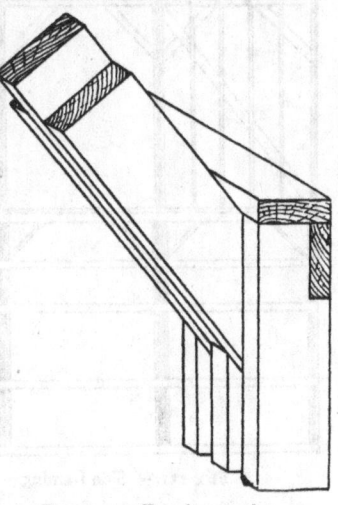

FIG. 110. — Framing at plate.

The two halves of the truss are bound together at the top by means of a short collar beam of 2 × 8 nailed to the two ridge braces. This collar beam should not be too long, for it must also support the carrier track, and if too long, will bring the carrier track so low as to interfere with the maximum filling of the mow.

It is very important that this type of self-supporting roof be designed to resist any side strain or racking which might result from high wind pressure upon the end of the barn. This is accomplished to some extent by the rigidity resulting from the roof sheathing, but additional bracing must be provided in the form of *sway bracing*, which ordinarily consists of long pieces of 2 × 8 fastened on diagonally beneath the rafters and firmly nailed at every

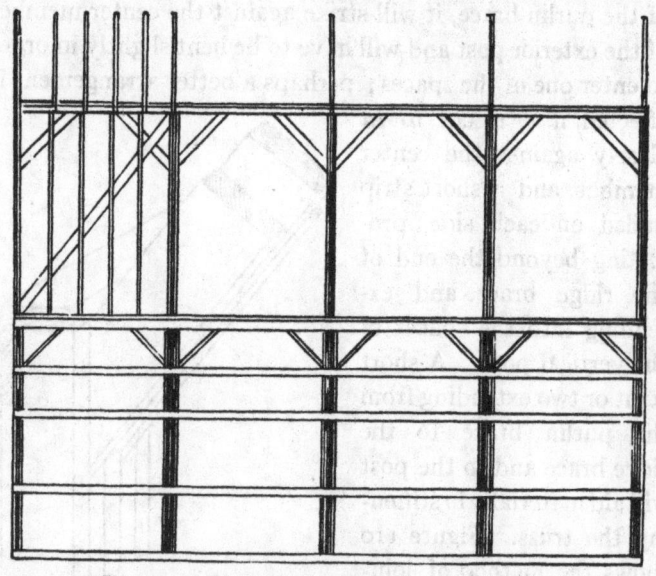

FIG. 111. — Side framing — framed for vertical siding.

joint. This is illustrated in Figure 111, a view of the side framing of a plank framed barn.

Figure 112 shows one method of arranging the planks in an end bent, studs being used to hold the horizontal siding. When, as is usually the case, a hay door is put in one gable, a vertical post may extend from the plate to the roof on each side of the door.

The plank frame has a number of advantages which make it especially desirable. Among these may be enumerated the following:

CONSTRUCTION OF FARM BUILDINGS

A saving is effected in the cost and amount of lumber used.

Timber can be used that could not otherwise be utilized.

A saving is effected not only in sawing, cutting, and hauling, but in time of construction as well.

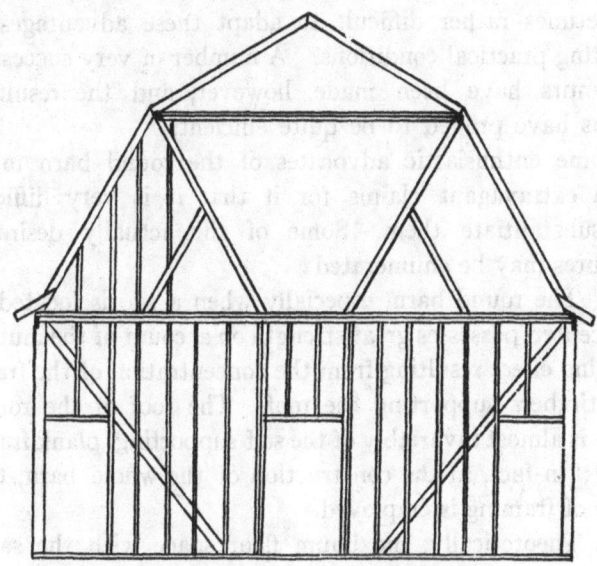

Fig. 112. — End bent — framed for horizontal siding.

Practically all interior timbers and braces are eliminated.

Full benefit is gotten of the self-supporting roof, combining triangles, long braces, and perpendicular timbers.

The possibility of weakening at the joints is eliminated.

A strong support for the hay carrier, with plenty of clearance, is provided.

The odds and ends of lumber can be utilized as braces.

THE ROUND BARN

Another development in the direction of economic building construction is the round barn, examples of which in a more or less modified form can be found in almost every locality. The round barn possesses some theoretical advantages which make its design very attractive, but it is sometimes rather difficult to adapt these advantages to existing practical conditions. A number of very successful attempts have been made, however, and the resulting barns have proved to be quite efficient.

Some enthusiastic advocates of the round barn make such extravagant claims for it that it is very difficult to substantiate them. Some of the actually desirable features may be enumerated:

1. The round barn, especially when a silo is located at the center, possesses great strength on account of the mutual bracing effect resulting from the concentration of the framing timbers supporting the roof. The roof of the round barn is almost invariably of the self-supporting, plank frame type; in fact, in the construction of the whole barn, this type of framing is employed.

2. Theoretically, maximum floor space with the same perimeter is obtained at a minimum of cost, since with the same perimeters in variously shaped figures, a circle gives the greatest area.

3. Increased storage space is provided because of the height of the roof necessary to give it proper support.

There may be other advantages of more or less degree of importance, depending upon the purpose for which the barn is used. It is in the interior arrangement of the floor devoted to stalls and bins that sometimes considerable difficulty is encountered. Unless the barn is very carefully planned, there is likely to be waste space and loss of efficiency in feeding and cleaning operations. The construction of round barns of large diameters is practically precluded by the inability to provide sufficient light when

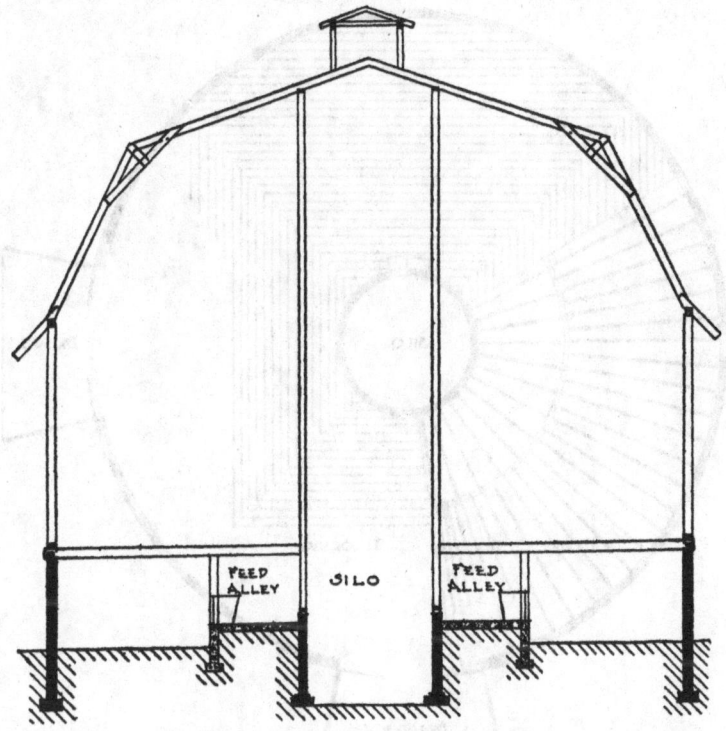

FIG. 113. — Framing of 60-foot round barn.

the interior stalls are located too far from the windows in the exterior wall.

Figure 113 shows the method of framing employed in a 60-foot round dairy barn. The plate, which is necessarily circular, is built up of six thicknesses of 1 × 6 on edge, the boards being put together so as to keep the joints well staggered; the purlin plate is constructed with a thickness equal to that of four boards. No posts are used, studs being used to support the walls and plate. Each pair of rafters are braced as shown in the illustration, and below

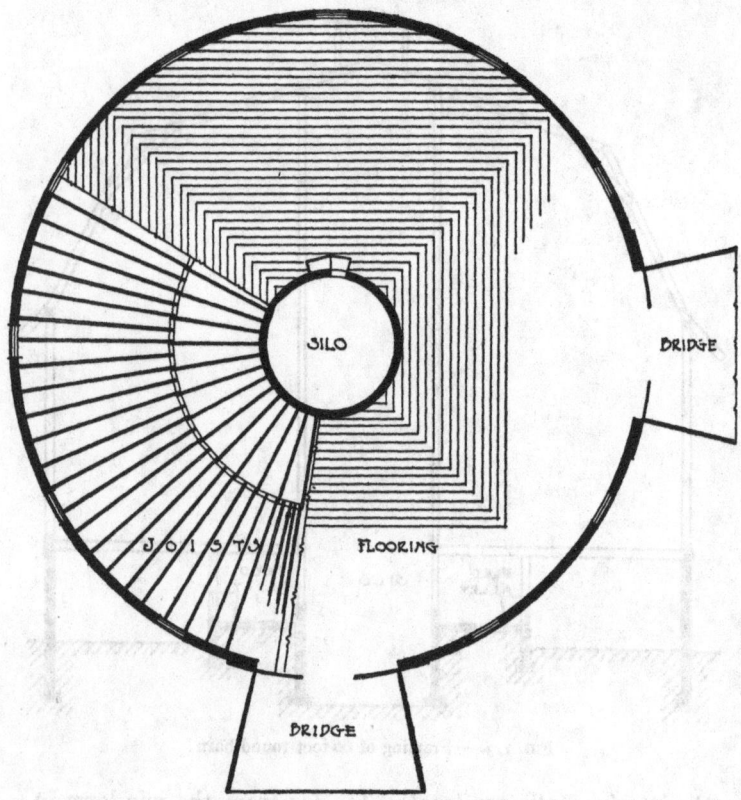

FIG. 114. — Floor joist arrangement.

the break in the roof an additional rafter is put in between each pair of regularly framed rafters, so that in the lower section of the roof there are twice as many rafters as in the upper section. The floor joists extend radially from the silo, as is shown in Figure 114, and have one interior support.

Dairy Barns

The importance attached to the proper construction and care of the dairy barn is emphasized in the following statement set forth by P. B. Tustin of the Health Department of Winnipeg: "The cow stable is the kitchen where the food for many city babies is prepared, and it is the duty of every farmer and dairyman to see that the kitchen is clean." The dairy barn should be kept as clean as a dwelling house, because the milk from the cows housed in the barn is consumed by humans, usually without any intervening converting or purifying processes, on the assumption that the milk as produced and handled is pure. That this assumption is generally incorrect is testified to by the fact that the dairy barns and cow stables on the great majority of farms are reeking with filth. In many cases the alleviation of this condition would be difficult and expensive, because the original design and construction was so very inefficient and unsanitary. A dairy barn can be constructed in such a way and of such materials as to permit of its being kept absolutely clean and sanitary at no great expense of time or labor.

The proper and economical erection of dairy barns involves great care and foresight in the design and arrangement in order to obtain the greatest efficiency. A barn is a rather expensive structure, and once built is not easily moved or altered in shape. The site is important; consideration must be given to the location as to the points of the compass, the position of the surrounding buildings, the proximity to the farm residence, and the appearance of the barn from the highway. The size should be suited to the amount of stock to be sheltered and the quantity of feed to be stored, and the interior should be so arranged as to facilitate feeding and caring for the stock. The appearance of the building when finished is a point to be given no small amount of attention, for the barn is a large structure and can easily be made to dominate the ensemble of the farmstead with a decidedly disagreeable effect. Some architectural features, such as a cupola, a cornice, and proper framing of windows and doors cost but little, yet do wonders in improving the appearance. Perhaps the most important feature of a dairy barn is cleanliness, and the use of concrete and steel where possible, and the installation of an effective system of ventilation will go a long way toward establishing this feature.

The arrangement of the interior of the dairy barn is a problem upon the solution of which there is a great difference of opinion. It seems that almost every dairyman has a different idea which he claims to be the best; this perhaps results from the special planning which each individual has had to do to satisfy the conditions of his own special case, and whether or not his solution is applicable to other equally special cases, is only problematical. The fact remains that in order to arrive at the best and most economical solution of the problem, the conditions existing on the farm upon which the barn is to be located and any special purposes for which the barn is to be employed must be carefully studied, and only general principles of arrangements can be considered here.

Broadly speaking, various arrangements of interiors resolve themselves into two kinds, namely: those in which the cows face towards the interior of the barn, and those in which they face the opposite direction. It is assumed that there will be two rows of cows in the barn, for long experience has shown that this gives the most practical results. The width of the barn will be determined by cer-

Fig. 115. — A modern dairy barn.

tain measurements of stalls, mangers, gutters, and alleyways that have been found by actual experience to be the most properly suited to the animals and their care.

As far as arrangement is concerned, the two methods given above differ only in feeding and cleaning facility. When the cows face in there is only one feedway, and two clean-out passageways; when the cows face out, there are two feedways and only one clean-out passage. Since generally the work of feeding is greater than that of cleaning, the work can be more easily and economically accomplished by having only the one feedway incident to the plan of having the cows stand with heads together. In fact this is the plan that is generally adopted, for besides the advantages mentioned heretofore, there are other points in its favor; the light falls on the rear of the cows, enabling the milker to see when the udders are clean and the stablemen to see better in cleaning out the stalls; there is less confusion in letting the cows in and out; the supporting posts can be placed in the line of the head rail, which is at the narrowest part of the cow, thus saving room; the ventilating system is usually at the walls of the barn, and the odor from the manure will not be so great as when the cows face out; and finally, it is easier to keep the barn clean when the slope of the floor is from the center to the outside, the drainage being more effective.

In any arrangement, the measurements of stalls, gutters, etc., are the same, and the dimensions given below are practically standard. The width of the manger will vary according to its construction from 2 feet to 3 feet, a wide, shallow manger being better than a deep, narrow one. The length of the "cow stand" or stall, from manger to gutter, should be 5 feet; this is a length which is suited to cows of all sizes, adjustment being made at the stanchion for short or long cows, so that all manure may be confined to the gutter and the cows kept clean. The width of the

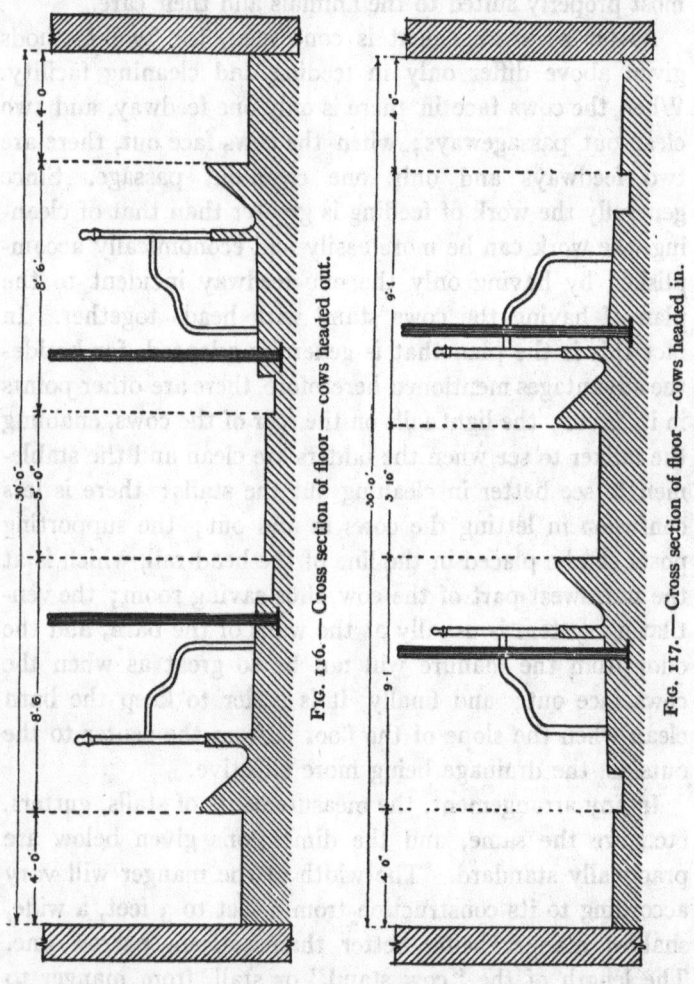

FIG. 116. — Cross section of floor — cows headed out.

FIG. 117. — Cross section of floor — cows headed in.

gutter should be 16 inches, and the depth of its bottom from the rear edge of the cow stand should be 6 or 7 inches;

the cow stand itself should have a slight slope towards the gutter. The passageway at the rear should not be less than 4 feet wide, and a width of 5 feet is amply large. The width of a central feedway need not be more than 6 feet between mangers, and in fact a narrower feedway is often used. When the cows are arranged heads out, the central passageway should be 8 feet wide if it is planned to be used as a driveway, but if a litter carrier is used, a width of 5 feet is sufficient. The width of stalls varies somewhat with the breed and size of cows, from 3 feet 2 inches to 4 feet; as a general average, a width of 3 feet 6 inches seems to be the best. Figures 116 and 117 in cross section, and Figures 118 and 119 in plan, illustrate the two arrangements discussed above, and show in detail what differences exist when the same plan is adapted to meet either requirement.

To further the effort to provide the most efficient sanitation of a dairy barn, the floors must be made of some material which is light and nonabsorbent, and which can be easily and thoroughly cleaned. Concrete floors seem to fill these requirements, though there is great objection to them, in northern regions especially, on account of their coldness. This can be obviated to a large extent by using plenty of bedding, or by putting in removable wooden platforms in the stalls during cold seasons. Concrete floors in barns are not to be troweled smooth, but finished with a rather rough surface with a wire brush or broom to eliminate the danger of the cattle slipping. There is on the market a floor brick made especially of finely ground cork and a special stiff grade of asphaltum, molded under pressure into brick form; they can be laid in cement mortar, but it is better to use asphalt as a binder. Whole floors made of asphaltum have been used, but they are

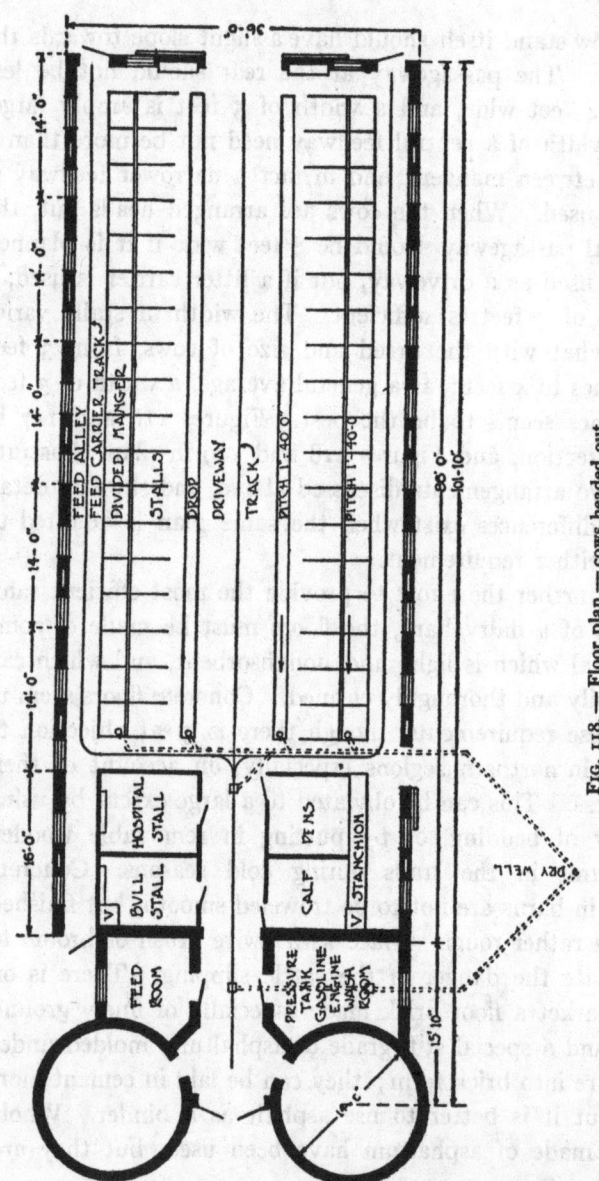

FIG. 118. — Floor plan — cows headed out.

CONSTRUCTION OF FARM BUILDINGS

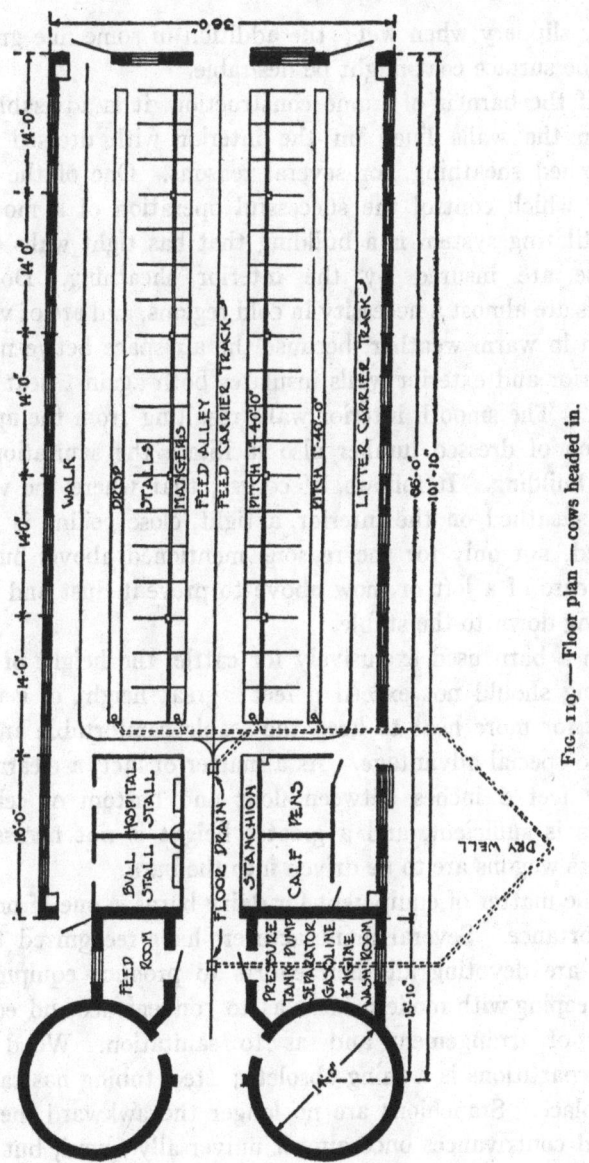

Fig. 119. — Floor plan — cows headed in.

very slippery when wet; the addition of some fine gravel to the surface coat might be desirable.

If the barn is of frame construction, it is advisable to have the walls lined on the interior with dressed and matched sheathing, for several reasons. One of the factors which control the successful operation of a modern ventilating system is a building that has tight walls, and these are insured by the interior sheathing. Double walls are almost a necessity in cold regions, and are of value even in warm weather, because the air space between the interior and exterior walls insulates both against heat and cold. The smooth interior walls resulting from the application of dressed lumber also facilitate the sanitation of the building. It follows, of course, that where the walls are sheathed on the interior, a tight, close ceiling is provided, not only for the reasons mentioned above, but in the case of a loft or mow above to prevent dust and dirt sifting down to the stable.

In a barn used exclusively for cattle, the height of the ceiling should not exceed 9 feet; great height of ceiling calls for more heat to keep the stable comfortable and is of no special advantage. As a matter of fact, a clearance of 7 feet 6 inches between floor and bottom of ceiling joists is sufficient, and a greater height is not necessary unless wagons are to be driven into the barn.

The matter of equipment for dairy barns is one of prime importance. Several manufacturers have recognized this, and are devoting all their efforts to produce equipment in keeping with modern ideas as to convenience and economy of arrangement and as to sanitation. Wood for stall partitions is a thing obsolete; steel tubing has taken its place. Stanchions are no longer the awkward, heavy wood contrivances once almost universally found, but are

CONSTRUCTION OF FARM BUILDINGS 243

made of light steel so arranged as to have a lateral swinging motion that gives the cow almost as much freedom as when outside, yet prevents her from moving backwards and forwards. Mangers are made of concrete or steel, the latter type being either fixed or movable. Even interior posts may be of steel, as shown in Figure 120. Both feed and litter carriers are part of the equipment of the dairy barn, and a proper arrangement of them permits of the carrying of ground feeds, grain, and silage to every manger in the barn, and the expeditious removal of all waste.

The gutters in the rear of the stalls should lead to a manure pit outside of the barn so that all the liquid manure can be saved and utilized. This pit may be constructed of concrete properly reënforced and waterproofed; it usually has to be put partly or wholly underground, depending on the floor level of the barn. If the liquid manure is to be applied in liquid form, a

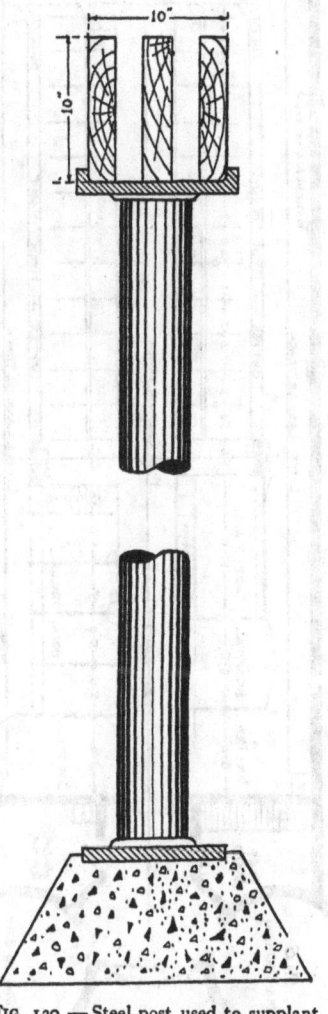

FIG. 120 — Steel post used to supplant wood construction.

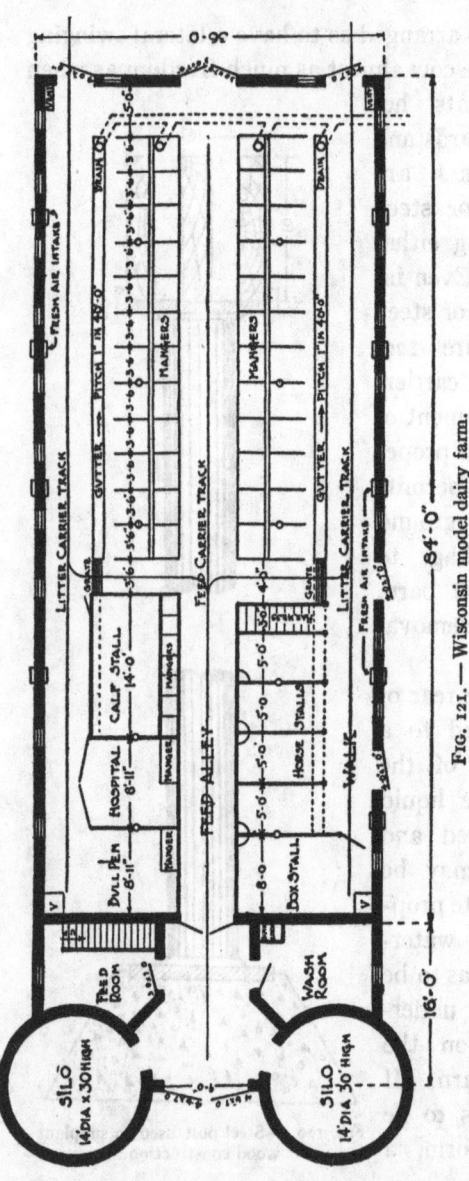

Fig. 121.—Wisconsin model dairy farm.

strong, serviceable pump should be located in the lowest part of the pit to pump it into tanks for transportation; otherwise a quantity of sawdust, leaves, tanbark, or similar substance can be put in the tank which will absorb the liquid and which can be handled with shovels.

Figure 121 illustrates the design selected by the State of Wisconsin as the model type of dairy barn for that state. The design was selected from numerous ones submitted in competition for a $1000 prize, the cost of the building not to exceed $2000. This was some years ago, however, and in all probability the barn will cost much more at the present time. It has a number of good features such as heading the cows in; providing stalls for calves, the bull, and sufficient horses to operate a dairy farm of 25 cows; a complete ventilating system; litter and feed carriers; excellent feeding arrangements.

Figure 122 is the floor plan of a specially well-designed large dairy barn housing more than sixty cows, eight horses, several calves, and a bull. Noticeable features of this barn are the exceedingly small amount of waste space, and the easy accessibility of the silos.

Where a number

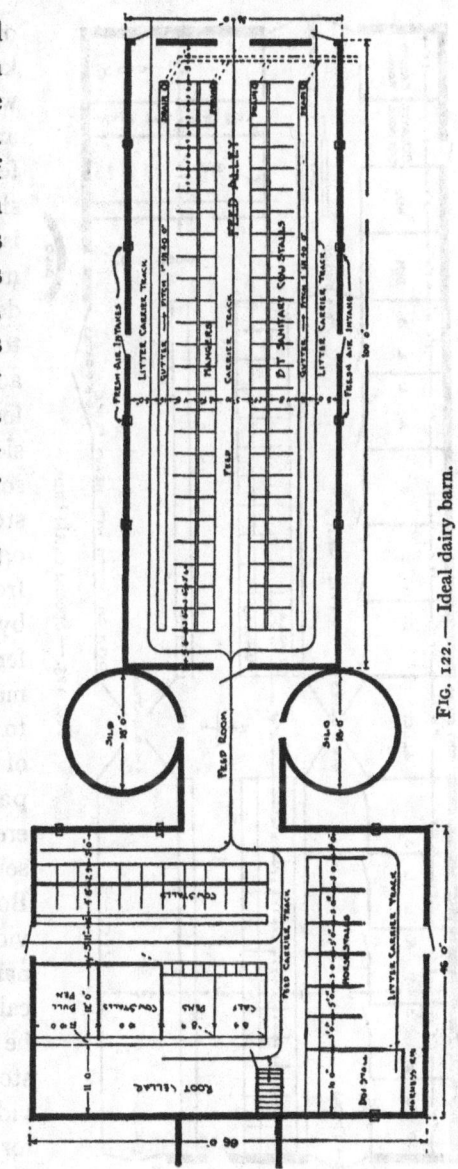

FIG. 122. — Ideal dairy barn.

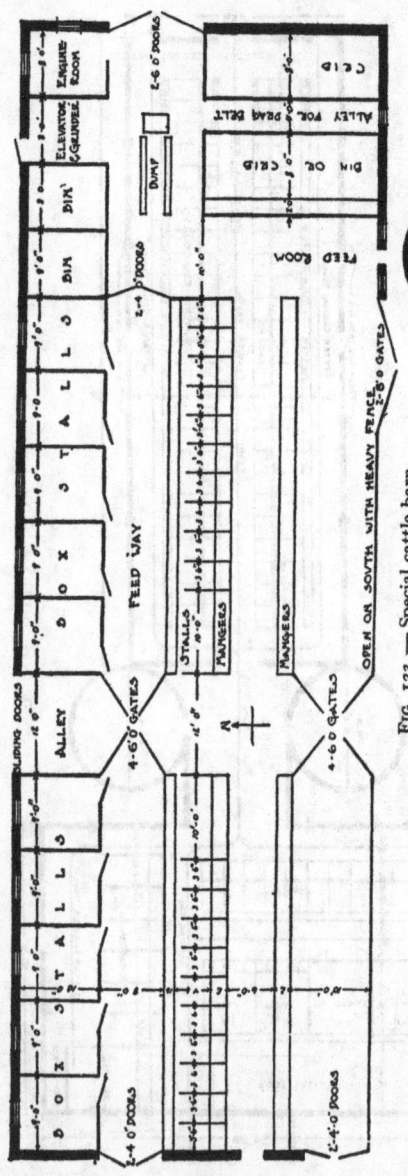

FIG. 123.—Special cattle barn.

of other cattle are kept in connection with the dairy herd and convenience in feeding them is desired, a barn such as is shown in Figure 123 may prove to be very desirable. This barn, to operate to the best advantage, should be located with the open side to the south; the south wall of the first story is left entirely open, being separated from the feed lot only by a heavy ordinary fence. In colder climates, if it is desired to keep the interior of the barn warm, a partition may be erected just at the south row of mangers. Box stalls are provided in which pregnant cows, cows with calves, or calves may be kept. Ample storage space is provided for feed, and for convenience in

CONSTRUCTION OF FARM BUILDINGS 247

handling an engine-driven dump an elevator may be installed. The driveway between the box stalls and single stalls is sufficiently wide to admit of a wagon being driven the entire length of the barn.

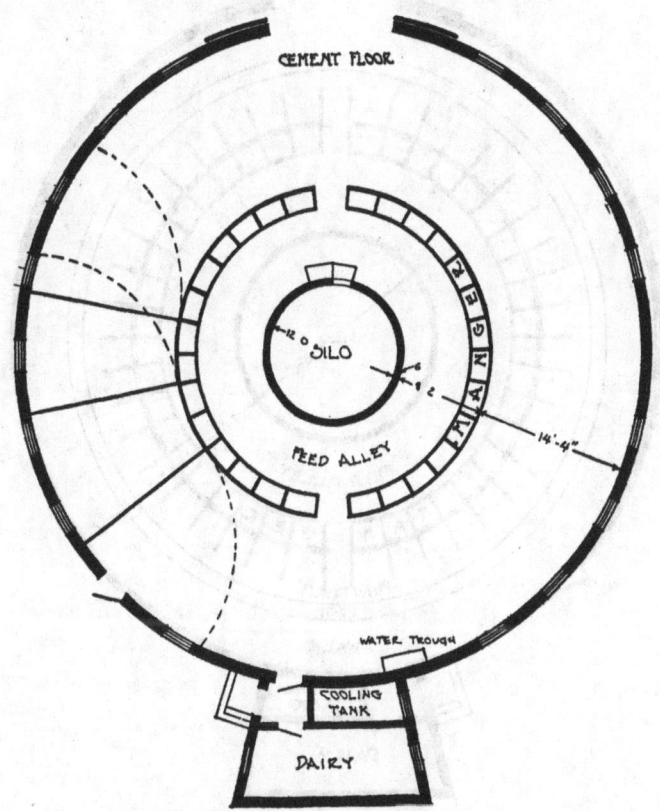

FIG. 124. — Round barn.   (Ill. Agr. Ex. Sta.)

Figure 124 shows in detail the somewhat novel arrangement adopted in a round barn at the Illinois Experiment Station. No stall partitions are used; the cows are simply fastened in the stanchions at feeding and milking

248 FARM STRUCTURES

time, being allowed the run of all the space outside the mangers the remainder of the time they are in the barn. Large box stalls can be formed, if necessary, by swinging

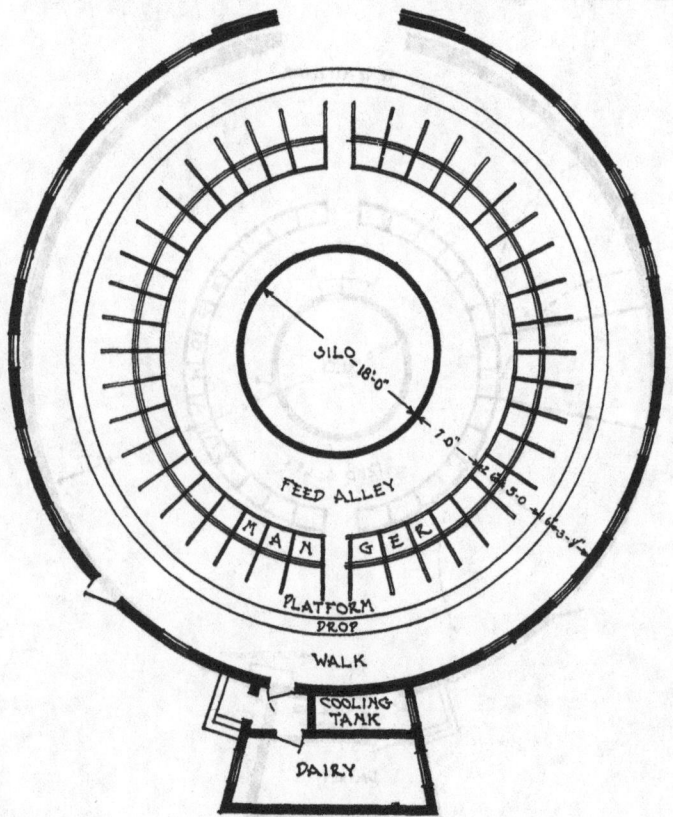

FIG. 125.— Illinois round barn — regular stalls.

around the large gates, as shown in the figure. The advantage of this particular arrangement lies in the fact that a large manure spreader can be driven in and around the entire building without the least difficulty. Figure

FIG. 126. — A successful round barn.

CONSTRUCTION OF FARM BUILDINGS

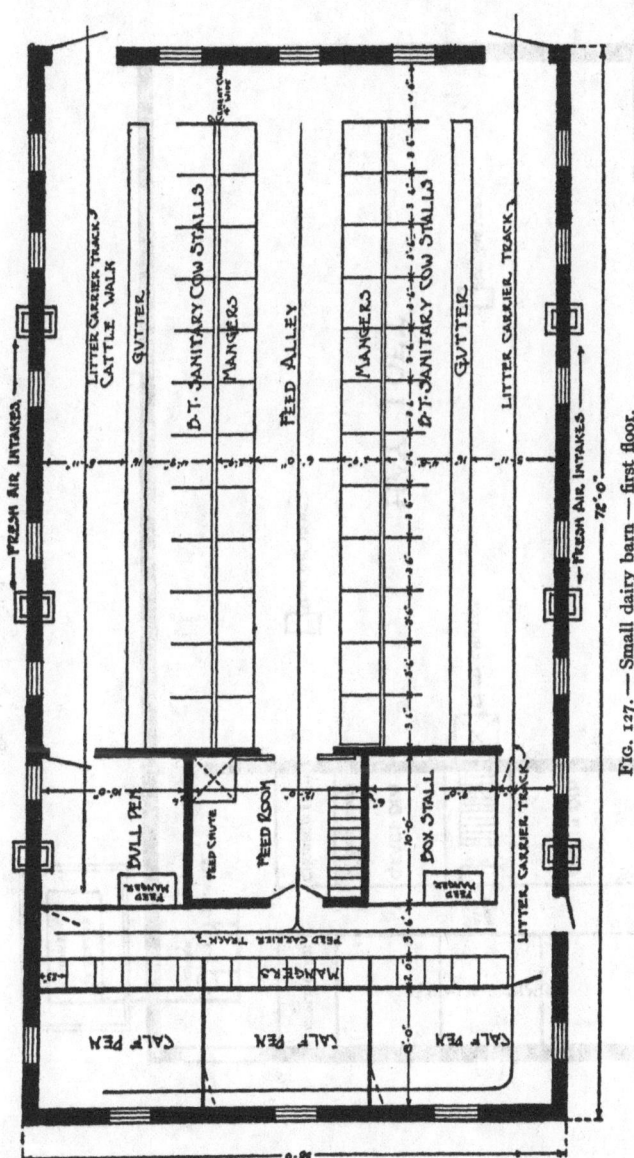

FIG. 127. — Small dairy barn — first floor.

250 FARM STRUCTURES

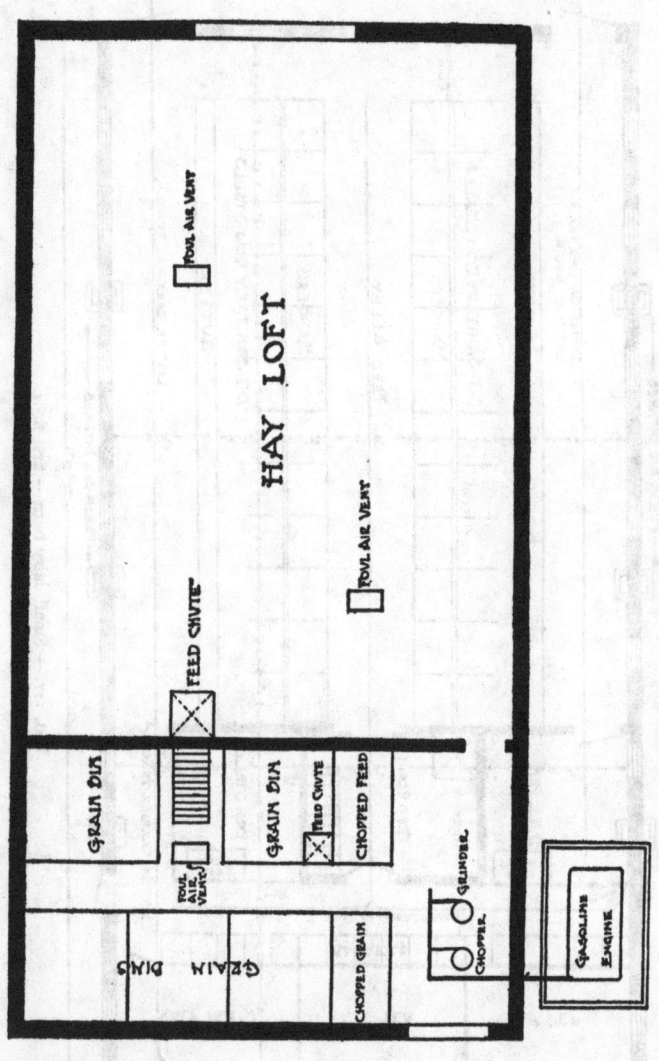

Fig. 128. — Small dairy barn — second floor.

CONSTRUCTION OF FARM BUILDINGS   251

125 shows the floor plan of this structure as it would appear were stall partitions introduced; in order to properly utilize the space, a larger silo, one whose diameter is 18 feet, must be built.

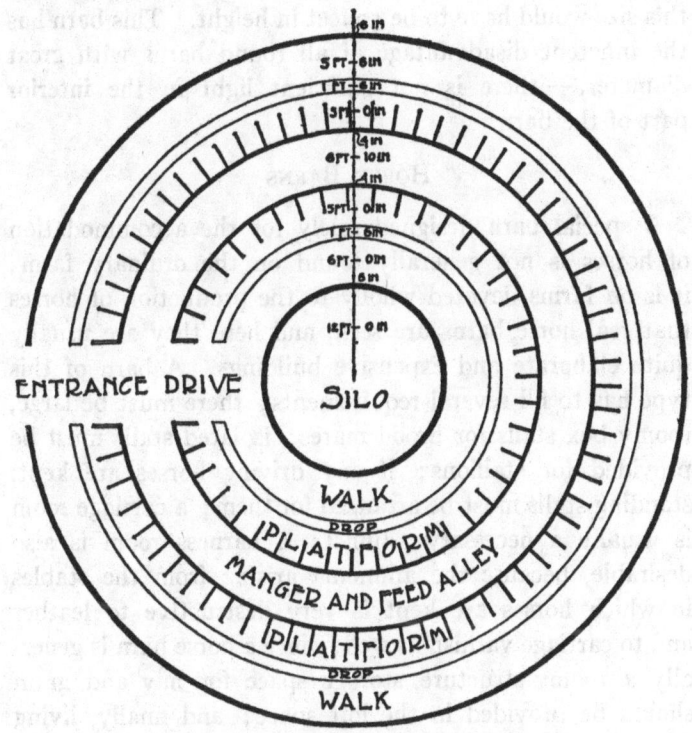

FIG. 129 — A large round dairy barn.

For a small dairy barn, the one whose floor plans are shown in Figures 127 and 128 is particularly good. The location of the grain bins, chopper, and grinder on the second floor admits of all the prepared grain being delivered to the first floor feed room, from which it can be expeditiously distributed.

Figure 129 gives the floor plan of a large round dairy barn. Close observation will show that at several points space is not utilized to the best advantage. The silo is 24 feet in diameter, and to reach to the roof of a barn of this size would have to be 50 feet in height. This barn has the inherent disadvantage of all round barns with great diameter, — there is not sufficient light in the interior part of the barn.

Horse Barns

A special barn designed solely for the accommodation of horses is not generally found on the ordinary farm; it is on farms devoted wholly to the production of horses that real horse barns are seen, and here they are usually quite elaborate and expensive buildings. A barn of this type has to fill several requirements; there must be large, roomy box stalls for brood mares; isolated stalls must be provided for stallions; if any driving horses are kept, standing stalls must be arranged for them; a carriage room is usually a necessary adjunct; a harness room is also desirable, because the ammonia arising from the stables in which horses are kept is very destructive to leather and to carriage varnish as well; since a horse barn is generally a roomy structure, storage space for hay and grain should be provided in the loft space; and finally, living quarters must usually be provided for the grooms and stablemen.

The character and temperament of horses are essentially different from that of any other farm animals, and this consideration must be kept in mind in horse barn construction. Horses are vigorous, active, and restless, and a greater solidity of structure than is necessary with other barns must be planned for. In box stalls the partitions must be very strong, especially the lower part; thus for a height of 5 or 6 feet they should be of 2-inch hard pine or oak, so that it cannot be broken or loosened by kicks; above this part should be a grating of ½-inch iron rods or heavy wire netting such as is used to protect exterior windows. The purpose of this netting is to keep the horse from being too closely confined, for otherwise he will become unusually restless and irritable, since he is a gregarious animal and resents deprivation of the company of his own kind. In some modern barns reënforced concrete partitions are meeting with favor, since they are sanitary, permanent, attractive, and offer no opportunity for gnawing or cribbing, a habit very common to young horses. Any sharp edges in either wood or concrete partitions should be carefully rounded off.

If a permanent manger is installed in the box stall, it should be bound with sheet iron so as to prevent gnawing. A portable box or manger can be provided, which is put into the stall only at feeding time, and removed when not in use; this, however, is more or less of an inconvenience. Some horsemen prefer to throw the hay on the floor of the stall, but this results in most cases in considerable waste, especially with long hay.

The size of box stalls varies; they should never be less than 8 feet in width, and a comfortable stall is 10 feet by 12 feet in size.

Standing stalls for single horses are usually about 5 feet wide, with a minimum of 4 feet 8 inches and a maximum of 5 feet 2 inches. The total length of a standing stall from front of manger to rear of passageway should be 14 feet, divided as follows: 2 feet for the width of the manger; 7 feet for the length of the horse stand; and 5 feet for the width of the passageway. These dimensions of course may be varied slightly to suit special conditions. The gutter at the rear of the horse stand for the disposal of the liquid manure should be at least 4 inches deep, and 16 or 18 inches wide; it can be left uncovered, but may be covered with heavy perforated cast-iron plates fitted into rabbets molded in the concrete floor.

The matter of floors for a horse barn is an important one. The floor is the part of the barn subjected to the hardest usage, consequently ability to resist severe wear is a prime requisite. A horse's feet are comparatively delicate, and the pawing and stamping characteristic of horses, if done on a hard floor, is likely to be injurious to them. With many horsemen a packed clay floor is the favorite, but this is insanitary and requires frequent repair. A wood floor of heavy plank is commonly put in barns, but it wears rapidly and there is danger of injury to the horses, should a plank break. A concrete floor seems to meet requirements best, but it has the objection of being very hard, so hard as to cause the feet of a horse to become tender when he has to stand upon it continually. A removable platform of 2 × 4 pieces, spaced $\frac{1}{2}$ inch apart and placed longitudinally in the stall, solves this difficulty, and it can be replaced at no great expense when worn through. The floor in the passageway should be roughened.

A sufficiency of light and adequate ventilation are two essentials of a good horse barn. Interior stalls, that is, stalls so far away from windows that good light does not reach them, are undesirable; 3 or 4 rows of stalls are sometimes placed in a barn, one row along each side wall, and one or two rows in the center, but it is better to have just the two rows of stalls along the outside walls and use the central portion as a place for exercising. Windows

FIG. 130. — A large round barn, showing double break in roof line.

should be plentiful and should be so arranged as to be easily opened; each one should be fitted with a wire screen to keep flies out during summer. In inclement weather ventilation should be accomplished by some such ventilating system as the King. Exterior doors to stalls and passageways should be made in two parts, the upper half to be replaced in summer time by a screened door, protected by heavy wire netting or by hardwood bars.

The provision for watering is unimportant as long as the water is pure. Inside water tanks are desirable, and if such are installed the plan of the barn must be such as will admit of facilitating the work of watering. Individual troughs for each stall are difficult to keep clean, and the

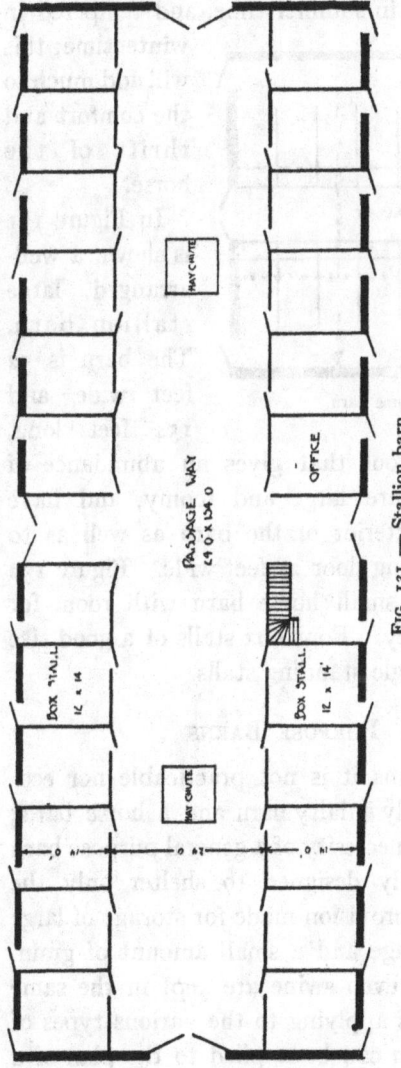

FIG. 131 — Stallion barn

256 FARM STRUCTURES

old-fashioned way of carrying water in buckets to the horses is a waste of time. The main requisites are that the water be fresh and cool in summer time, and tempered in winter time; this will add much to the comfort and thrift of the horse.

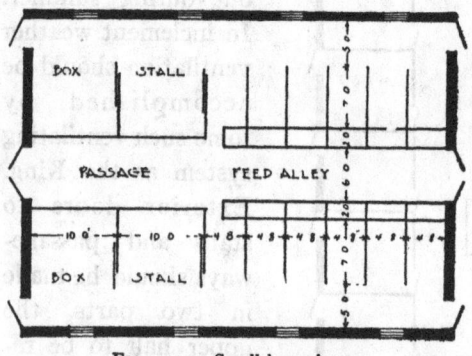

Fig. 132. — Small horse barn.

In Figure 131 is shown a well-arranged large stallion barn. The barn is 52 feet wide and 154 feet long, with a self-supporting roof that gives an abundance of loft room. The stalls are large and roomy, and have doors opening to the exterior of the barn as well as to the large interior exercising floor 24 feet wide. Figure 132 illustrates a convenient small horse barn with room for 18 horses in an emergency. Four box stalls of a good size are provided and ten single standing stalls.

General Purpose Barns

On a great many farms it is not practicable nor economical to have separately a dairy barn and a horse barn; on farms of this kind the necessity of a general purpose barn is obvious. It is usually designed to shelter only the cows and horses, with a provision made for storage of large quantities of hay or forage and a small amount of grain, but sometimes sheep or even swine are kept in the same building. The principles applying to the various types of barns as heretofore given can be applied to the plan of a general purpose barn, and an economical and attractive structure can be arranged.

The horse stalls and cow stalls should, if possible, be placed on opposite sides of the building, on account of the difference in the amount of space required. Three cow stalls require only as much width of space as two horse stalls. The grain bins can be placed on the second floor if necessary, and this arrangement is especially practicable when a portable grain elevator is available with which to place the grain in the bins. The purpose of the grain bins is not so much to provide storage for a large amount of grain, as to make easily available some grain during inclement weather when it would be an annoyance to have to carry it in from an exterior separate crib. A harness room should be located at some point convenient to the horse stalls; it may be fitted with harness racks or with harness hooks, and may serve as a repository for medicines, etc., which have no other special place.

THE FARM RESIDENCE

The actual work of building, of putting materials together so as to make a finished structure, is the work of the contractor or builder; the preparation of the plans for the contractor to follow, and the decision as to the kind of materials to be employed so that a durable and economical structure may result, is the province of the architect; but the collection and correlation of ideas and features relating to houses so that the structure when finished may be a home to suit his requirements — *that* is the privilege and pleasure of the owner.

In a previous chapter structural details have been considered fully enough to enable the student of them to become sufficiently well acquainted with building operations to supervise construction and be able to differentiate good construction from bad. Rarely does the farmer attempt the construction of his residence himself unless he has had some training in carpentry, for this kind of work requires considerable skill and experience to be accomplished economically. While a man with little experience may succeed well with a poultry house, a granary, or even a simple barn, the construction of a residence involves so many comparatively difficult operations that should he attempt it, the result would be an unsatisfactory piece of work, besides being a very expensive one.

Occasionally, and quite commonly, in fact, the whole proposition of building the house is put into the hands of a contractor. He is told to put up a house with so many rooms, to cost not more than a certain sum; and with these bare directions he proceeds, realizing that in all probability the size of the completed structure will be the only thing the owner will consider; absolutely no thought is given to the location of the house with reference to the most beautiful vistas, to arrangement of the interior for the convenience and comfort of its inhabitants, nor to any development of natural resources in the way of attractive surroundings.

For this reason the employment of an architect is almost a necessity. True, the work of inexperienced owners sometimes results admirably; but this is more to be assigned to the probability of their being real artists, than to any real skill in house design and construction. As a rule, the houses bearing the earmarks of an owner's design are better than those planned by builders. Many house owners, and their wives and daughters, have very clever ideas about building, and when these ideas are correlated and incorporated into a design by some one who is skillful in such a line, a charming house usually results. Many people are of the opinion that the employment of an architect is a useless expenditure of money; as Helen Binkerd Young states in her admirable discussion of the farmhouse: "Few persons believe that they have no right to build until professional help can be afforded; yet such a position would be well taken. Houses stand not for a month nor for a year, but for a generation; by them the thrift of a community is judged, by them the ideals and taste of a community are formed. He who deliberately builds an ugly house condemns himself as a poor citizen; while he who builds a beautiful house proves himself a good citizen, for his personal effort contributes to the public welfare."

The planning of any house is a serious undertaking, and the special conditions surrounding the problem of farmhouse design and the peculiar requirements to which a farmhouse is subjected make the planning of the farmhouse a task worthy of long and careful study. Where considerable time can be given to the thought of the design, much better results will follow; and in the majority of instances this is possible, for rarely is it necessary to build a farmhouse in a hurry. From the time the idea of a new house originates, the owner and his family should be on the lookout for ideas that can be incorporated in the design. Other houses should be visited, and arrangements for convenience, comfort, and attractiveness should be noticed; do not be afraid to copy good features — the best architects copy freely. The more good features that can be included in a design, even if they are not original, the better the design will be. A file should be kept of all ideas accumulated, and should be given to the architect when the actual preparation of plans is begun. This is what the architect wants; his desire is to please his patrons, and the more ideas of theirs he has that he can incorporate in the plans, the better he will do it. Leave the development of the style and the utilization of location, materials, and ideas to the architect, but be sure to give him the benefit of what you have learned to be worth while.

The general problem of designing a farmhouse cannot best be solved unless the whole farmstead and the surrounding topographical features are considered; the farmhouse is merely a single unit in the general farm scheme that should unite into one workable system lands, barns, and dwelling so that permanent economy may result. Organized farming and organized housekeeping are two essentials of successful agriculture, and no element contributes more to this success than a well-arranged farmstead. Hit and miss methods of construction, causing a continual round of tearing down, reconstruction, and makeshift, result in waste of time, money, and labor, and interfere seriously with the efficient prosecution of farm operations.

In the beginning of the plan of the farmhouse, cognizance must be taken, then, of the fact that it will be more difficult to plan than either a city or a suburban residence; for it must not only be a home, but it must fill the place of business houses and outside markets which supplement the city home. To conform to these needs, it must of necessity have a comparatively large floor area, in order that provision may be made for a business center and for store space; and the larger the floor area to be utilized, the greater will be the opportunity for the occurrence of mistakes in planning.

Architectural Styles

It is difficult to discuss the style of architecture employed in farmhouses, for as a matter of fact the true American type has as yet not been evolved. Of course, in the design of a house it is not necessary to copy any style that has gone before, for it is entirely possible to make a design that will conform in certain features to several styles, or it may conform to none, yet in either case it will have charm. "Style is not a mere external covering, something to be applied outside. Style is vital — structural — as well as ornamental." Much depends on individual style, and the materials that can be obtained at the least expense, for often the limit of the appropriation is a very strong controlling factor.

The Colonial style, either pure or more or less modified, dominates the houses of eastern United States to a very great degree, and there are numerous examples of it scattered over the whole country. It is an interesting style, brought to America by early colonists from England. Colonial houses are usually broad, with a hall in the center and rooms on both sides, the parlor and living room on one side, dining room and kitchen on the other, with the bed chambers arranged symmetrically on both sides of the second-floor hall. Some of the distinguishing external characteristics of the Colonial style are tall, stately columns, small porches, roofs either gambrel or high-gabled, narrow eaves, and symmetrically placed windows. This style of architecture is sometimes peculiarly adaptable to the farmhouse, for it requires considerable room to admit of its proper development, and building sites are ample in the country, if nowhere else. However, care must be exercised to maintain the simplicity and stateliness of the Colonial house, or it quickly loses its charm. Some modern adaptations of the Colonial style have departed so widely from the original that but few points of similarity remain.

It is interesting to note how differently the development of the English country house in the last few centuries differs from the American. Both derive their inspiration from the same source, — the Early Georgian, — but when the Colonial and the English house are placed side by side, the difference is very wide indeed. Apparently, our English cousins have very carefully considered the value of the site of the house, for in no other land are the houses in more perfect harmony with their surroundings; they appear to be indigenous to the soil, as all good houses should appear, and nestle down in the midst of trees and flowers as if some master gardener had there planted them.

Modern English houses are quite solidly built, thanks to stringent modern English building laws. The framework of older English houses was of timber, the walls being filled in with brick masonry instead of being covered with shingles or clapboards. Since timber became scarce and more valuable, the builders have evolved an architecture of brick and stone quite as attractive as the older houses. In either case the English house is easily recognizable by its charm of design, long-sloping, graceful roof, usually broken by a few gables, and small projecting wings or ells which serve as a location for stores. The English type is entirely practical for American farmhouses, and with a few modifications fits remarkably well into certain landscapes, particularly where the country is rolling and wooded.

Another style occasionally met with on eastern farms is the modern Dutch; it was originally brought from Holland by the early Dutch settlers. Houses built in this style are usually quite sedate and symmetrical, the quaint irregularity of the English house being absent; they are placed with a central entrance on a broad side to the front, with broad porches supported by large, simple columns. In these houses the living room is quite large and is often used as a dining room. Bed chambers are almost always on the second floor.

A distinct American style of architecture is that, common in the West, known as "Mission." The influence of Spain is plainly shown in its development. The Mission style succeeds best in locations where there is an abundance of land available, which should be quite decidedly rolling. In California, where are seen some of the best examples of Mission architecture, the picturesque mountains and hills form an excellent setting which seems difficult to obtain in any other region. The features of the Mission style are large, plain wall surfaces, which are usually of stucco, occasional high, severe towers, and close grouping of windows. To brighten up the dull, dead surface of the cement walls, red roofing tile and bright color in window sash and frames are used.

The California bungalow illustrates another style which is an adaptation of the Mission style to suit modern building methods, materials, and conditions. The outstanding features of the bungalow are low construction, broad, overhanging eaves, and comparatively low roofs; a true bungalow is of but one story, though there are many houses called bungalows which are of one and one half or even two-story construction. To be the most satisfactory, the bungalow should be left in its native land; transported to the plains of the Middle West or to the rugged hills of the East, it loses some of its charm.

A type of architecture that appears to be particularly well suited to rural surroundings is that which has had its origin in the Middle West, and is variously known as the "Chicago," or "Natural," or "American" style. It has traces of the influence of several styles that have preceded it, and has even a touch which gives a "Japanese" effect. The houses built in this style have a peculiar quiet dignity and an appearance of solidity. Low-pitched roofs, wide eaves, high windows, and originality in interior planning combine to make a distinct impression of attractive simplicity which harmonizes well with the country landscape.

The Interior of the Modern Farmhouse

Convenient and satisfactory interiors to suit all the conditions of city houses are common, but the arrangement of farmhouse interiors has been given little attention. This is unfortunate, because the problem of domestic help is much more serious on the farm than in the city; in the majority of instances the country housewife has a number of other duties besides that of the care of the house, and every effort should be made to produce as convenient an interior as possible. Efficient housekeeping is just as great an essential of successful agriculture as is efficient farming, and where the farmer feels that a well-arranged group of farm buildings is a requisite for efficient operation, his wife is entitled to have a workshop equally efficient.

The Kitchen. — To the housewife, the kitchen is the most important part of the house; much of her time is spent there, at one task or another, and the kitchen arrangement must be such that it will the most satisfactorily conserve the housewife's energy.

The tendency among modern farmhouses is to reduce the size of the kitchen. Houses built in times gone by were built to serve different purposes and to fit different circumstances, and often the kitchen had to serve also as dining room. For a small family this might serve now, but evolution in farm life has emphasized the desirability of a separate dining room; this allows of a more satisfactory solution of the problem of taking care of an extraordinary number of men on special occasions, such as in threshing, ensilage cutting, etc. This problem exists to a greater or less extent in every farming community. The actual size of the kitchen may, of course, be varied a little, but the floor dimensions should be such that their product approximates 150 feet, with no dimension less than 9 feet. Square kitchens can be more efficiently arranged than rectangular ones, since any point can be reached with a minimum amount of travel. The location of a kitchen should preferably be on any side other than the south, since sunshine, which should be utilized for the living room, is for the most part of the year not highly desirable for kitchens; a morning or evening exposure to the sun is sufficient. Good lighting and good ventilation, however, are two kitchen essentials, and can best be accomplished by locating the windows high enough to admit of table space below them. Doors should be well placed and as few in number as possible, since they occupy wall space, and wall space is valuable for cupboard and shelving. The question of a pantry is a debatable one; some housewives cannot do without one; others much prefer cupboards; however, when a sufficiency of cupboard room can be provided, it is probably better to eliminate the pantry, especially when other storage space is supplied.

Regarding the general arrangement, a few principles must be kept in mind. Cleanliness is the first requisite; this necessitates simplicity in arrangement, and accessibility of the various articles of furniture, such as range, sink, etc. Where the dining room is separate, the kitchen and the kitchen processes must be hidden from view from the dining room as much as possible. Tables, cabinets, and similar articles must be so placed as to be most convenient, and so that no unnecessary energy is spent in moving about a great deal. The sink should be constructed so that there is no possibility of any contamination; most sinks are placed too low, an increase of six inches in the height from the floor being advantageous.

Dining Room. — In a farmhouse the dining room often serves the additional purpose of living room. In any case, cheer and comfort are associated with it, and its design and arrangement should emphasize these attributes. The room should be large enough to permit of the placing of a good-sized table, with sufficient room around the seating of the table to allow the easy passage about that is required in service. Since an ordinary table is $4\frac{1}{2}$ feet wide, and the width taken up by persons seated at the table will increase the necessary table space to at least 7 feet, it is evident that 11 feet is the minimum dimension for a dining room.

Lighting is an exceedingly important detail. The strongest light should be located so as to shine directly into the eyes of as few persons as possible. It is better to have several small windows advantageously placed than one or two large ones; it is a good idea to have the windows placed rather high in the wall, since this arrangement gives a light that is less glaring, and the space below the windows can be used for the placing of the dining room furniture. The ideal location for a dining room is in an eastern exposure, so the morning sun may send in its cheer and brightness for the morning meal; if in a western exposure, in summer time the rays of the evening sun may interfere with the comfort of the family when gathered together at dinner.

FIG. 133. — A typical horse barn.

A butler's, or pass pantry, is a feature of dining room arrangement, the desirability of which is much to be doubted. The formality which often accompanies meal time in urban life is absent in the country, and a single swinging door for passage to and from the kitchen is better than a pantry. It is usually possible to provide some sort of a pass slide between the kitchen and dining room, and this, with a movable table mounted on smooth-running casters, will simplify the serving of meals.

Living Room. — In olden times there was perhaps greater attention given to the living room than any room in the house; that the living room was used for dining purposes was merely an incidental feature. Later this type of arrangement was changed, and the "parlor" was added, in which the more ostentatious element of the social phase of life was presumed to be taken care of. The parlor as a room was not a success, except that it supplied a single, seldom-needed want — that of an orderly place in which to entertain the unexpected formal caller. Fortunately, modern evolution has changed this condition of affairs, and the old-fashioned cold, cheerless, stiffly formal and uncomfortable room has been transformed into the modern living room, a room in which comfort for the family is the primary consideration.

The best exposure for a living room is one to the south, so that full benefit of sunshine may be had in winter time. An abundance of light should be provided; a living room is generally rather long, and to adequately light it will require a number of windows. The interior arrangement admits of wonderful opportunities for special features. A fireplace is almost essential. Built-in bookcases and window seats add marvelously to the comfortable appearance of the room, as do wall paneling and beam ceilings.

The room should be furnished and used for the purpose as indicated by its name, and should be the center of the social life of the family.

Bedrooms. — Inconvenience is certain to arise if the farmhouse is not furnished with a sufficient number of bedrooms. Four is the minimum that should be provided, and sometimes more than that number are necessary, especially at times when there are farm operations requiring more than the usual quota of men. One bedroom may well be placed on the first floor, though many good designers insist that all bedrooms should be on the upper floor. Ample ventilation is of course an essential, and for this reason every bedroom should have at least two exterior walls with windows in them, to provide a cross draft. Built-in wardrobes are desirable, and usually cheaper and more satisfactory than separate pieces of furniture.

Other Features. — Due to the peculiar conditions of farm life, there are certain features which ordinarily must be included in the design. One of these features is a room in which the men of the farm, when coming in from work, may wash and leave their working clothes. This room should, of course, be located in an inconspicuous position at the rear of the house, and access from it to the dining room should be had by means of an easy and convenient passage. The furnishings should include a lavatory, an ample supply of clothes racks, and perhaps a strong, simple bench seat. The wash room may be made of fair size and may then serve as a laundry room, and in some localities, where wood is a fuel, as a wood storage room.

Roomy, comfortable verandas are pleasant additions to a residence in any locality; their location will, of course, depend somewhat upon the style of architecture adopted, but in no case should they be placed where the maximum of use cannot be gotten out of them. So often the only porch the farmhouse possesses is one at the back that is so covered with boxes, washing machines, and what not, that there is no place for a chair, or a stiff, formal front porch that is always exposed to the hot sun. A wide, convenient veranda, protected from the sun by shade trees, climbing vines, or even awnings, and furnished with comfortable porch furniture, is never a matter of useless extravagance; it is a real necessity.

A bathroom is more of a necessity in a farmhouse than it is in a city house. No modern farmhouse is without its system of water supply under pressure, and this renders the installation of the bathroom fixtures a comparatively simple matter. Where possible, two bathrooms should be provided, one for the members of the family, and another for employees. This, of course, will mean additional expense, but it may be entirely desirable in some instances.

In the accompanying design of a convenient farmhouse, Figure 134, an attempt has been made to incorporate as many desirable features as possible. The arrangement has been made with the idea in mind of fulfilling certain local conditions, a highway passing the house to the south, giving the house a south front. Of course, modifications of this plan, or an entirely new one, would be necessary to satisfy different conditions.

Beginning at the front of the house, we have first a large, roomy veranda, on the west side of the house, so that benefit may be obtained of any winds that may blow during the summer. Entering the house, we find a small vestibule, with doors leading into the office, and into the living room; the business caller may be quickly ushered into the office, or the guest may enter directly into the living room. The office is really but an alcove off the

270   FARM STRUCTURES

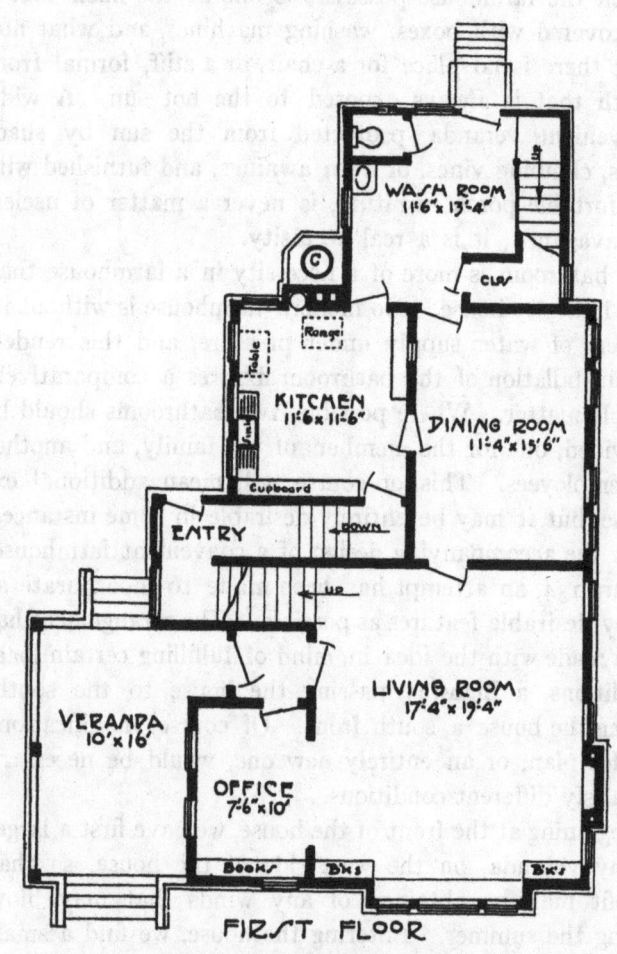

Fig. 134.—A convenient farmhouse.

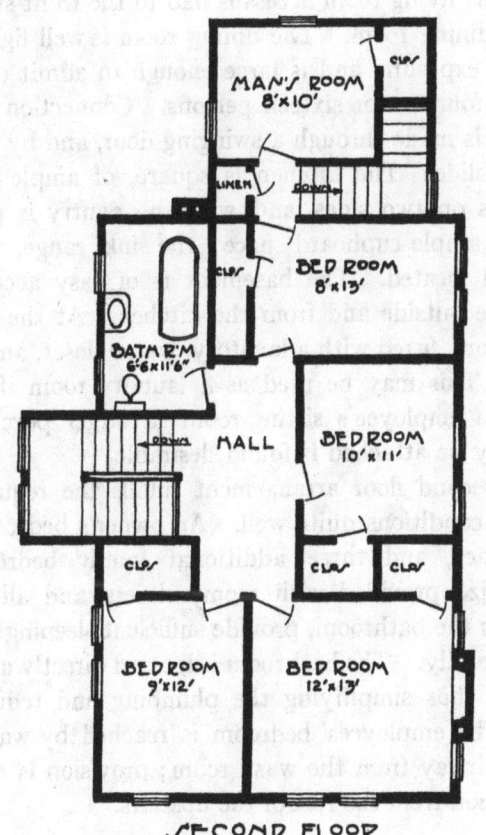

FIG. 134.—A convenient farmhouse.

living room, which in itself is a large, bright room, with fireplace, window groups, bookcases, and built-in window seat. The living room is particularly fortunately situated, with exposures in three directions, east, south, and west. From the living room access is had to the front stairs, and to the dining room. The dining room is well lighted, has an east exposure, and is large enough to admit of a table seating fourteen or sixteen persons. Connection with the kitchen is made through a swinging door, and by means of a pass slide. The kitchen is square, of ample size, has windows on two sides, and, while no pantry is provided, there is ample cupboard space; the sink, range, and table are well located. The basement is of easy access, both from the outside and from the kitchen. At the rear is a wash room, fitted with a lavatory, water closet, and clothes closet. This may be used as a laundry room if desired, or as an employee's sitting room; a larger porch at the rear may be attached if found desirable.

The second floor arrangement fulfills the requirements of farm conditions quite well. An owner's bedroom, with a fireplace, and three additional family bedrooms, of ample size, provided with roomy closets, and all of easy access to the bathroom, provide sufficient sleeping quarters for the family. The bathroom is located directly above the kitchen, thus simplifying the plumbing and reducing its cost. The employees' bedroom is reached by way of the back stairway from the wash room; provision is made for its isolation from the rest of the upstairs.

CHAPTER VI

VENTILATION

THE provision of some means for ventilating churches, factories, school buildings, large office buildings, etc., is an exceedingly important factor in the design of every such structure, and illustrates the importance attached to the necessity of a constant supply of fresh air; it may be that this necessity is recognized in these cases because one cannot help but notice that in places which are more or less crowded the ill effects resulting from such crowding are especially emphasized. Yet an ordinary residence which has the design of a good ventilating system incorporated in its plans is an exception; were it not that common residence construction is so inferior as to provide a sort of a ventilation through loose windows and doors, the inhabitants of the house would be constantly breathing vitiated air. As for farm buildings in which animals are housed, the possession of a ventilating system is of such extreme rarity as to be almost an oddity.

A more thorough appreciation of the value and importance of fresh air as a means of improving and maintaining animal health is bringing about an improvement in methods of modern building construction. Numerous devices are available whereby the ventilation of residences can be effected, and progressive farmers are making use to a greater extent than ever of ventilating systems for their stables.

Ventilation includes not only the supplying of fresh air, but the removal of vitiated air as well. Since the results of ventilation are satisfactory only when the air of the inclosure is constantly maintained at such a degree of purity that it can be breathed without harm, it naturally follows that the ventilating system will work at its greatest efficiency when both of the aforementioned requirements are met. If the first requirement only is fulfilled, there results simply a dilution of the air already present; the percentage of harmful constituents is constantly increasing, however, so that, conditions remaining the same, a point is soon reached beyond which an increase in the supply of fresh air is useless, since no vitiated air is being removed.

From the above it is seen that there are two main reasons why ventilation is rendered a necessity: fresh air must be supplied to furnish oxygen, the constituent of air essential to life; the harmful gases, such as carbon dioxid, ammonia, and methane, must be removed. To these reasons there may well be added a third one, that of removal of bacteria which thrive in ill-ventilated rooms.

Residence Ventilation

The ventilation of ordinary residences is usually accomplished, if at all, in conjunction with the action of the heating system; in fact, the introduction of pure air can only be accomplished properly in connection with the heating system, and any system of heating that does not make provision for an adequate supply of pure air is incomplete and imperfect.

One of the primary facts we must know in the design of a ventilating system is the amount of air needed; this can be very readily determined, if only we know the amount of carbon dioxid given off in the process of respiration. From

VENTILATION

400 to 500 cubic inches of air per minute are required for each adult person, on the basis of 20 to 25 respirations per minute, and assuming that the average person requires 20 cubic inches at each respiration. The following is a comparison of inspired and expired air, showing the approximate effect of respiration:

	Inspired Air	Expired Air
Oxygen, per cent of volume	20.26	16
Nitrogen, per cent of volume	78.00	75
Moisture, per cent of volume	1.70	5
Carbon dioxid, per cent of volume	.04	4

The last-named element is usually taken as an index of the purity of air, since the greatest variation occurs with this constituent, there being one hundred times as much of carbon dioxid in expired as in inspired air. Various degrees of purity, that is, various percentages of carbon dioxid, may be taken as standards, depending upon conditions. To determine the amount of air necessary, let

C = cubic feet of air required per person per minute.
a = parts per 10,000 of carbon dioxid in expired air.
b = cubic feet of air breathed per minute.
n = parts of carbon dioxid permissible in inspired air.

Then
$$C = \frac{a\,b}{(n-4)}.$$

If $a = 400$, and $b = \frac{1}{3}$, the formula becomes

$$C = \frac{133}{(n-4)},$$

and by substituting various values of n, we can determine the amount of air required per person for any standard of purity. The following table is derived in this way:

Permissible Number of Parts of CO_2 per 10,000 Parts of Air	Cubic Feet of Air Required per Person	
	Per Minute	Per Hour
5	133	800
6	67	4000
7	44	2667
8	33	2000
9	27	1600
10	22	1333
12	17	1000
13	13	800
16	11	667
18	9.5	571
20	8.3	500

Authorities differ greatly as to the degree of purity of the air required, but a common and accepted practice is to adopt 8 or 9 parts of carbon dioxid per 10,000 of air, making necessary about 30 cubic feet of air per minute, as a minimum.

When the rooms for which the ventilating calculations are to be made are heated with gas, special provision has to be made on account of the contamination of air resulting from the combustion of the gas. Usually the combustion of one cubic foot per hour of gas vitiates air to the same extent as is done by one person, and a gas burner ordinarily consumes 4 or 5 cubic feet per hour, so that an allowance of air equivalent to the requirements of that number of persons must be made for each gas burner. Candles, lamps, etc., also consume some air, and provision should be made for them.

The determination of the size of air flues can be made if two factors are known; namely, the total amount of air required per unit of time, and the distance the air in flues will flow in the same unit of time. Professor Carpenter, in his treatise on "Heating and Ventilating Buildings,"

VENTILATION

states that for residences the air velocity in flues is likely to be according to the following table:

	Velocity in Feet per Second	
	Inlet Duct	Outlet Duct
First story	2.5 to 4	6
Second story	5	5
Third story	6	4
Attic	7	3

Suppose then that the size of ducts is required for a third-floor room in which 8 persons are located, and where the standard of purity is to be maintained at 9 parts of carbon dioxid per 10,000 of air.

Cubic feet of air required per hour: $1600 \times 8 = 12800$
Velocity of air in inlet duct: $6 \times 3600 = 21600$
Velocity of air in outlet duct: $4 \times 3600 = 14400$
Cross section area of inlet duct: $\frac{12800}{21600}$ = .6 sq. ft. or 86 sq. in.
Cross section of outlet duct: $\frac{12800}{14400}$ = .9 sq. ft. or 130 sq. in.

As a matter of fact, however, separate ventilating flues are not usually installed in residences. The cracks existing in loosely fitted windows and doors are usually counted upon to supply exits for the vitiated air; that these are often inadequate is testified to by the offensive odor that greets one's nostrils and the sensation of stifling that one feels upon entering a crowded room. In a residence heated with hot air, excellent ventilation is furnished if the air to be heated is taken through the supply duct from the exterior of the building. If, however, the supply duct leads from the interior of the building, there is little real ventilation, but simply a continuous circulation of air already more or less vitiated.

Many residences heated by steam or hot air are furnished with a good supply of fresh air by the employment of a device known as the "indirect radiator." This is simply a radiator placed in a flue leading from the exterior of the building to the room to be heated and ventilated. As the air within the flue is heated, it rises to the opening within the room, this starting a current that is continually bringing a supply of fresh, warm air into the room.

The presence of a fireplace or an open grate sometimes serves to provide a satisfactory method of ventilating a room. Special fireplaces of patented construction are available which act in much the same manner as do the indirect radiators, by drawing in a supply of cold air from the outside, heating it, and passing it out into the room. Such a contrivance is illustrated in Figure 135.

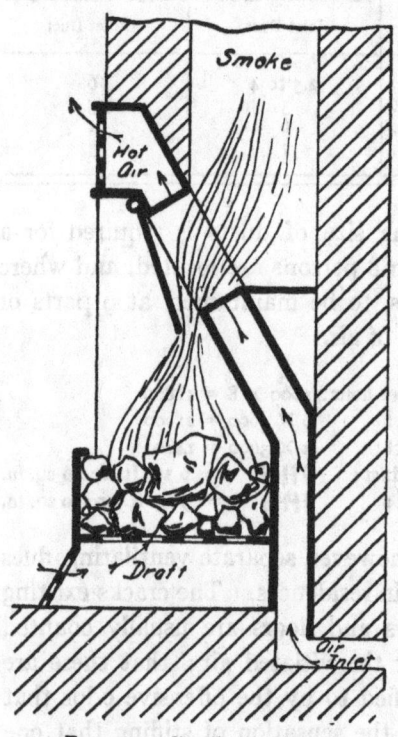

FIG. 135. — Ventilating fireplace.

Sometimes the question arises of ventilating rural schoolhouses and other stove-heated buildings of a similar type in which gatherings of people occur. In cases of this kind

VENTILATION

the adoption of some plan whereby the ventilation can be accomplished without the production of any noticeable drafts and any marked decrease in the temperature of the room is necessary. The smoke flue should be arranged to rise directly up from the stove, as shown in Figure 136.

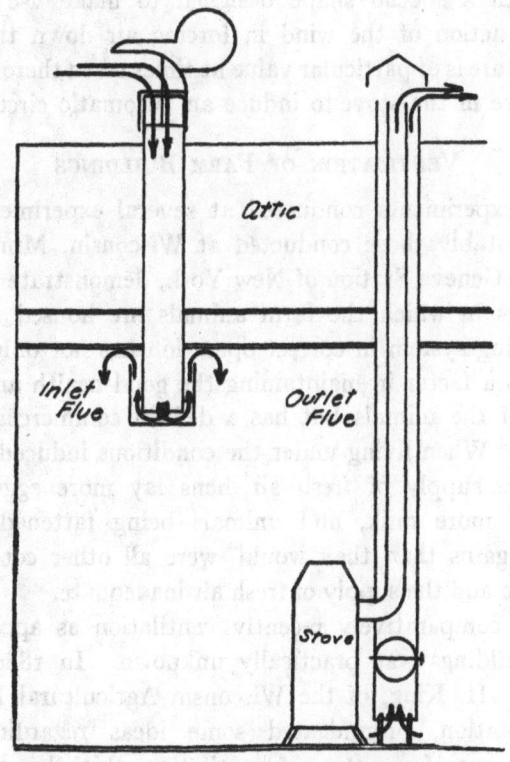

FIG. 136. — Ventilating system for stove-heated room.

Surrounding the smoke flue, and reaching within a foot of the floor, is the outlet flue; its diameter may usually be 18 or 20 inches, and it should extend to the same height as the smoke flue, with a cap at the top and a damper at the bottom with which to control the amount of air admitted. Some sort of an inlet flue must be provided; a good type is shown in the figure. It consists of a vertical flue with its lower outlet protected by a distributing cap; the upper end has a revolving cap fitted with a vane and with a special shape designed to make use of the driving action of the wind in forcing air down the flue. The feature is of particular value at times when there is little or no fire in the stove to induce an automatic circulation.

Ventilation of Farm Buildings

The experiments conducted at several experiment stations, notably those conducted at Wisconsin, Minnesota, and the Geneva Station of New York, demonstrate that in buildings in which the farm animals are housed a good ventilating system in correct operation has not only great value as a factor in maintaining the good health and condition of the animals, but has a definite commercial value as well. When living under the conditions induced by an adequate supply of fresh air, hens lay more eggs, cows produce more milk, and animals being fattened make greater gains than they would, were all other conditions the same and the supply of fresh air inadequate.

Until comparatively recently ventilation as applied to farm buildings was practically unknown. In 1889, Professor F. H. King, of the Wisconsin Agricultural Experiment Station, promulgated some ideas regarding the development of a system of ventilation which has become almost universally known as the "King" system. It is, in reality, a "natural" system, one in which a few natural elementary principles are applied; but since the application of the principles is especially ingenious and shows the result of much care and thought, it is entirely appropriate that the system should be given the name of the man who has done so much toward the development of a really important feature of farm building construction.

The King ventilating system consists of two sets of flues, one for the removal of impure air, the other for the inlet of a supply of fresh air. Air that has been breathed contains a high percentage of carbon dioxid and is heavier than pure air; consequently, the outlet flues should begin near the floor of the building, where the impure air will collect. When this air is removed, fresh air is drawn in through the inlet flues which open near the ceiling; the purpose of this is to afford it an opportunity, when necessary, to become warmed, since greater heat is likely to be found at this point.

The employment of any artificial means of producing a positive circulation of air through the ventilating system is usually impracticable, consequently we must make use of some or all of the following natural causes:

1. Difference in temperature between the air within the stable and that without. When air becomes warmed, as it will in a building occupied by animals, it expands, decreases in density, and rises, thus making way for colder, heavier air.

2. Wind pressure on the windward side of a building tending to force air into it.

3. Suction on leeward side of a building due to aspiratory effect of wind.

4. Aspiration at the top of the outlet flue.

The second and third methods are perhaps not of much moment except when the wind is rather high; the last-named method is the really important one. The diminished pressure existing at the top of the outlet flue as the result of wind blowing across it is a source of positive air circulation, the action being similar to that of certain spray machines or atomizers in which fluids are aspirated out of receptacles by passing swift currents of air over their orifices.

To design a ventilating system of this type, all that is necessary to know is the velocity of air through the flues, and the amount of air required by the animals housed in the building for which the ventilating system is being designed. As a matter of fact, the first quantity is exceedingly variable, sometimes reaching 500 feet per minute, but if we assume a rate of 300 feet per minute, we can err only on the safe side. Professor King gives the following table of approximate air requirements for various farm animals:

>Horses — 70 cubic feet per minute per head.
>Cows — 60 cubic feet per minute per head.
>Swine — 23 cubic feet per minute per head.
>Sheep — 15 cubic feet per minute per head.
>Hens — 0.5 cubic feet per minute per head.

Then to ascertain the cross section in square feet of the flue that will supply sufficient air for any number of animals, we simply divide the total number of cubic feet of air required by 300. For example: required, the size of outlet flue for a general purpose barn accommodating 12 horses and 4 cows.

$$12 \text{ times } 70 = 840$$
$$4 \text{ times } 60 = 240$$
$$1080$$

$\frac{1080}{300} = 3.6$ square feet = cross section area of flue.

We should make two outlet flues, perhaps, one 1 foot by 2 feet, the other 1 foot by 1½ feet.

The outlet flues should be constructed so as to be as tight as possible, either of double thicknesses of wood, with building paper between, or of galvanized iron. In general the latter will be found to be slightly lower in cost. The location of the outlet flues will of course be governed somewhat by the shape and arrangement of the barn, but the chief precautions to observe are that the flue be constructed solidly and that it have as few bends as possible; a few sharp bends will be sufficient to destroy the air current within the flue.

The inlet flues should be individually quite small, about 6 by 12 inches in cross section, but their total cross-section

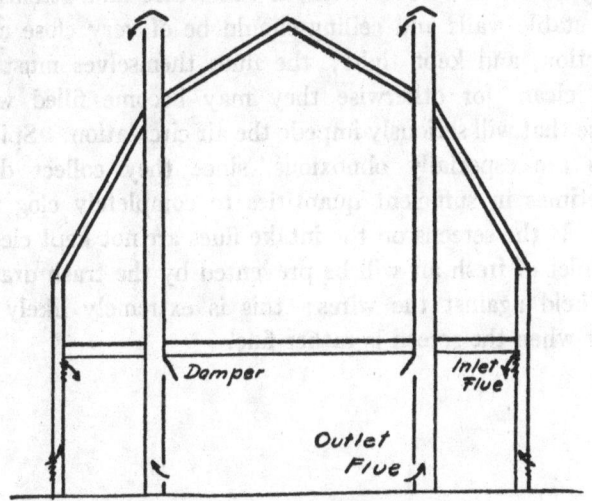

Fig. 137. — King ventilating system.

area should be slightly in excess of the total carrying capacity of the outlet flues, to insure plenty of fresh air. As with the outlet flues, their location will depend upon the plan of the barn, but generally they are placed in the exterior wall of the barn, not more than 12 feet apart. They should open into the barn near the ceiling, and their exterior opening should be at least 3 feet below their interior opening, to prevent them from acting as outlet flues. Their construction is similar to that of the outlet flues, but often when a barn is being built the intake flues can be made self-contained in the wall, of vitrified sewer tile, galvanized iron, etc. It is well to protect both the interior and exterior openings with coarse screen, to prevent birds from nesting inside of them.

This system, which is illustrated in Figure 137, will operate with uniform success, if it is borne in mind that no system will operate without some care and attention. The stable walls and ceiling should be of very close construction, and kept tight; the flues themselves must be kept clean, for otherwise they may become filled with refuse that will seriously impede the air circulation. Spider webs are especially obnoxious, since they collect dust sometimes in sufficient quantities to completely clog the flue. If the screens on the intake flues are not kept clean, the inlet of fresh air will be prevented by the trash drawn and held against the wires; this is extremely likely to occur when the screen is rather fine.

CHAPTER VII

LIGHTING FARM BUILDINGS

For centuries candles constituted the only source of artificial illumination, and even to-day their convenience and adaptability make their use (under certain limiting conditions) highly practical. The discovery of the enormous fields of petroleum in the United States, the distillation of it to produce kerosene in quantities, and the development of the wick type of kerosene lamps marked another epoch in the progress of illumination, and so simple and cheap in operation are these lamps that probably more than ninety per cent of the farmhouses of this country are lighted by them. The development of the isolated gas and electric lighting plants, however, marks still another step which is as far in advance of the kerosene lamp as the kerosene lamp is in advance of candles, or even farther.

The three great questions of importance in considering lighting systems are as follows: economy, or the question of cost of equivalent illumination; sanitation, bearing on health and efficiency or illness and inefficiency; and the æsthetic consideration, the pleasure, the attractiveness of fine illumination, that adds cheer and charm to the evening hours in the home.

CANDLES

In some homes candles are used to a certain extent because lamps or other forms of artificial illumination are disliked on æsthetic, or, in some cases, ostensibly on hygienic grounds. Speaking broadly, illumination by means of candles is either very inadequate so far as ordinary living rooms are concerned, or, if adequate, is quite expensive. Experiments have shown that the degree of illumination does not increase in nearly the same proportion as does the size of the candle; that "sixes" are nearly as efficient, as regards the amount of light, as "eights" or "twelves." The amount of light derived from an ordinary candle is slightly in excess of that emitted by the standard candle, so that to obtain an equivalent illumination of 100 candle power requires only 85 or 90 ordinary wax or paraffine candles. But, actually, the essential objects in the room could be as efficiently illuminated by perhaps 30 or 35 candles, properly distributed so as to concentrate the light where desired, as by 2 or 3 gas burners, or 4 or 5 kerosene lamps. With sources such as the latter, the illumination is of a much greater intensity near the source than is necessary. In this respect candles have an advantage over other forms of lighting, and, when considered on this basis, compare favorably in cost of equivalent illumination.

KEROSENE LAMPS

Kerosene lamps are so common that a discussion of them is almost unnecessary. The essentials of the lamp are a reservoir for oil, a burner, a wick for carrying the oil from the reservoir to the burner and constituting a part of the burner itself, and a shade for the protection of the flame from drafts. The burner is so constructed that the wick can be raised or lowered, thus controlling the amount of wick projecting above the sheath; the end of the wick, to which the oil is carried by capillary action, holds the flame, and the greater the portion of wick projecting, the higher will be the flame. Air to supply the flame is carried in through perforations in the lower part of the burner.

In lamps of ordinary size the candle power developed varies from 5 to 25, depending upon the purity and nature of the oil, upon the size and shape of the wicks, and upon the height of the flame. The cost of illumination by this method approximates that of acetylene lighting, and is about one third that of candles. In spite of the wide use of the wick type of the kerosene lamp, it is not an especially good form of illumination, since its light is yellow and not restful, and the products of the combustion cause an odor that is quickly perceptible unless the room is well ventilated. When a flat wick is used, the intensity of the light from the lamp is generally unequal in different directions, less light emanating from the edges of the flame than from the sides. In a flat acetylene flame this same difference in intensity exists, but to so small a degree as to be practically negligible.

Air Gas Lamps

The system of lighting by so-called air gas used for raising mantles to incandescence in upturned or inverted burners is a somewhat recent development, though the method of producing air gas has been known for years. "Air gas" is ordinary atmospheric air, more or less completely saturated with the vapor of some volatile oil, which saturation results from passing the air over the oil; if the oil is highly volatile, no heating is necessary to produce the required saturation, but for a less volatile one, gentle heating is advisable.

Though air gas has been available for many years, its use in flat-flame burners was not at all satisfactory, and it was not until the advent of the incandescent burner that it could be used advantageously for illuminating purposes. Various systems using gasoline, alcohol, and even kerosene, are on the market, and operate with varying degrees of success. Since it is very difficult to control the exact composition of the gas, there is great likelihood of variability in the amount of light emitted. The quality of the light will remain practically constant where incandescent burners are used, since in this case it is from the glowing particles of ceria, thoria, or similar metallic oxides that the light is derived.

Portable, self-contained lamps with incandescent burners using gasoline, alcohol, or kerosene are sold, and give fair illumination when once in operation; but it is sometimes quite difficult to get the generation of the right quality of gas started, especially when the oil is of low grade. From the nature of the construction of the lamp, it is also difficult to keep the mantle from breaking, so the maintenance cost is rather high.

Acetylene Lighting

Acetylene is a gas of which the most important application at the present time is for illuminating purposes, for which its properties render it especially well adapted. The light of a bare acetylene flame resembles sunlight very closely in composition or "color," it being more nearly a pure white light than any other common light used for illuminating purposes. Acetylene lighting presents also certain important hygienic advantages over other forms of lighting, in that it exhausts, vitiates, and heats the air of a room to a less extent, for a given yield of light, than do either coal gas, oils, or candles.

Acetylene is made by the interaction of water with a solid substance called carbide of calcium, or calcium carbide; all that is necessary is to bring the two into contact within a suitable closed space. A diagrammatic representation of the simplest form of an acetylene generator is shown in Figure 138. It consists of a closed vessel containing water in the lower part and an arrangement for holding carbide in the upper part so that a regulated flow of the carbide into the water will occur. Immediately the generation of the gas will ensue, and the gas thus produced is led away through the distributing system of pipes to the burners.

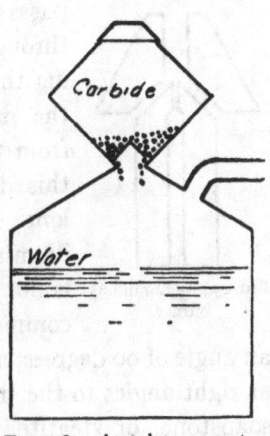

Fig. 138.—Acetylene generator.

The method above described, that of carbide-to-water generation, is the one most commonly used in acetylene generators. Water-to-carbide generators are manufactured, but are not so satisfactory as the first-named type. For portable lamps, both table or stand lamps and vehicle lamps, the water-to-carbide system of generation is more desirable, since it can be more easily and definitely controlled under the rather hard usage to which portable lamps are subjected.

The burners used in an acetylene lighting system are of two general types, the luminous and the incandescent. An "incandescent" burner is one in which the fuel burns with a flame which is in itself atmospheric or non-luminous, the light being produced by causing that flame to play upon some extraneous refractory material that has the property of emitting much light when raised to a sufficiently high temperature. A "luminous" burner is one in which the fuel is permitted to combine with oxygen in such a way that one or more of the constituents of the gas evolves light as it undergoes combustion.

U

With the luminous burner some means of cooling it to prevent ultimate destruction is necessary. For this reason, luminous burners are constructed upon the principle shown in Figure 139; the gas rushing out through the central passage injects a certain amount of air through the side passages, thus surrounding the gas with a thin coating of air, and the mixture is burned a short distance from the top orifice. One tip only of this description evidently will produce a long, slender, jetlike flame, in which the illuminating power of the acetylene flame is not developed economically, so that in common practice two tips are located at an angle of 90 degrees, as in Figure 140, yielding a flat flame at right angles to the triangle. These burners are made of soapstone, or steatite.

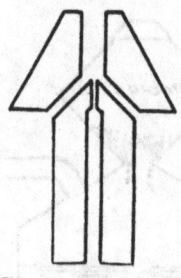

FIG 139.—Luminous burner.

To operate an incandescent burner with success, the gas must be pure, and be supplied under an even, steady pressure. The burner itself consists of a mixing tube with adjustable air inlets some distance back from the orifice, over which the mantle is hung, the whole being surrounded by a glass or mica shield. A gas mantle consists of a mesh of combustible material, such as cotton or ramie fiber, which has been impregnated with solutions containing certain "rare earths," such as thoria, ceria, etc. When used, it is adjusted to the burner, then ignited, and the combustible mesh is consumed, leaving a skeleton composed of the substances with which the mesh was impregnated.

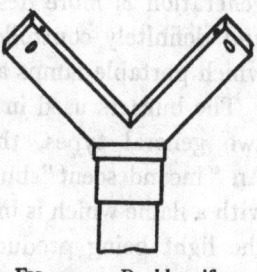

FIG. 140 — Double orifice burner.

The best globes that can be used for acetylene lights, and this applies to any other kind of light as well, are those made from some material which protects the eye from the bright and direct rays of light, yet disperses and diffuses the light so that none of it is lost, but all is used for illuminating. Plain white glass, unless the surface is specially shaped in prismatic form, is quite unsatisfactory for globes. Colored or tinted globes should not be used where the highest light economy is wanted, though this is often sacrificed for effect.

Considerable prejudice exists against acetylene because of the fatal explosions that have occurred in residences where lighting systems have been installed. The explosions have been caused by the bringing of a flame into a chamber in which there had been a leakage of the gas. By installing the acetylene plant in a chamber separate from the house, and employing reasonable precautions against possible danger, an entirely satisfactory degree of safety can be secured.

Electric Lighting

Electric lighting is an especially attractive method of illumination, because, with the use of the modern high efficiency lamps, the cost is not great, and it is safe and convenient. With properly arranged circuits the light is instantly at one's command, and no groping in the dark is necessary to find it. This method of illumination is especially advantageous to the farmer, in that it permits lighting not only the residence, but the barn and other buildings of inflammable character in a safe and efficient manner.

An installation simply to supply illumination may be made, but, where it can be afforded, a larger system in which part of the installation can be used for other purposes and which is of sufficient size to supply considerable power, is advisable. There are a number of small machines, particularly about the house, that can be so easily, economically, and conveniently operated by electricity, that the use of it can hardy be dispensed with. Storage batteries are conveniently used in connection with the generator, in order that power may be available whenever the generator is not running. These are particularly desirable in private plants for lighting, for sufficient battery capacity can easily be provided to supply power for fans, sewing machines, etc., so that these may be run at any time when the generator is not in use.

In the design of an electric lighting system, the first thing to ascertain is the number of lamp hours required. This is done by finding the total number of hours all the lamps are to burn. From this we can determine the number of battery cells to use, since each cell will furnish an average pressure of a voltage of about 2 volts. Hence, if 110 volt lamps are used, 55 cells will be required, — probably more than this, since a cell when nearly discharged will give only 1.8 volts. Battery cells, however, are quite expensive, and by using lamps of lower voltage, say 25 or 30, the number of cells can be reduced to about 15. One lamp permits one ampere of current to flow, so the capacity of the battery in ampere hours is equivalent to the number of lamp hours.

When a battery cell is fully charged, it will give a pressure of about 2.6 volts, so that the entire battery will give a pressure of approximately 39 volts. Since in charging a battery the current must flow into the battery in a direction opposite to the flow of the current when the battery is discharging, the entire voltage of the battery is opposed to that of the generator; that is, the battery is connected to the generator so that it tries to drive current through

LIGHTING FARM BUILDINGS

the generator while the generator is driving its current at the same time into the same end of the storage battery. Thus, to enable the generator to charge the battery, it should be able to generate a greater current than the battery; and in charging the battery 8 or 9 amperes may be used, though 5 amperes may be the normal rate, so the current delivered by the generator must be at least this amount. Generators are rated by the kilowatts of energy they produce. The number of kilowatts of energy produced is equal to the product of the voltage and amperage divided by 1000. In ordinary installation the voltage is about 45 and the amperage is 9, so the kilowatts produced is approximately one half.

Most isolated lighting plants are driven by means of gasoline engines. While theoretically the power of the generator and engine should be about equal, practical considerations, such as high ratings of the engines and the fact that their full power is not developed unless they are properly adjusted, make it advisable to have an engine with 50 per cent greater power than that required to drive the generator. For a $\frac{1}{2}$-kilowatt generator, a 2 horse power is not too large.

A switchboard and apparatus with which to control the generator and storage battery is the next consideration. The switchboard itself may be either slate or marble, the latter being much more costly. The switchboard equipment will include the following:

1. Rheostat, to control the voltage of the generator.
2. Ammeter, to measure the amount of current.
3. Voltmeter, to measure the pressure.
4. Circuit breaker, to disconnect the battery and generator in case of overload or reversal of direction flow of current.

5. End-cell switch, to control the voltage of the battery so that it may be kept practically constant.

6. Plug switch, to admit of different connections to the voltmeter so that the voltage may be measured at several places.

7. Two main switches, to connect the generator, battery, and lamp circuit in any desired manner.

A diagram of the wiring is shown in Figure 141. It will be seen that S_1 is a double throw switch, by which the

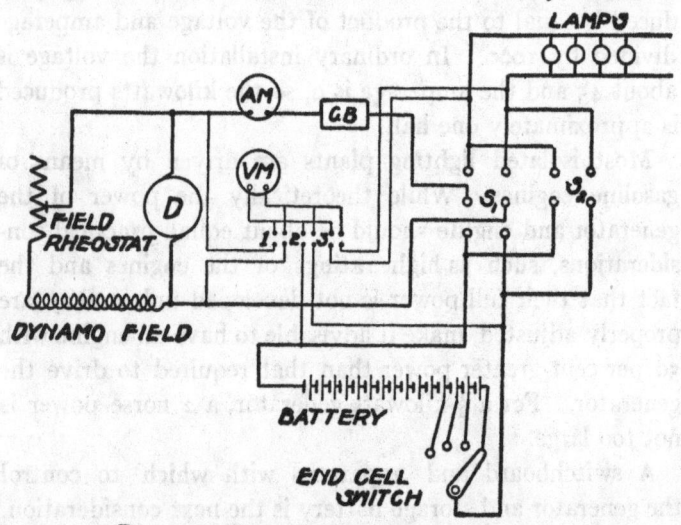

Fig. 141. — Wiring diagram for electric installation.

generator can be connected to the battery for charging, or it can be thrown over so the lights can be operated directly from the generator; in the case of the latter, the generator field rheostat must be adjusted to reduce the voltage to about 26 or 27 volts, or else the lamps will soon be burned out by the excess voltage. By leaving this switch open and closing S_2, the generator circuit is opened and the battery is operating the lights.

The size of the wire to be used will depend upon the amperage, and all wires should be large enough to carry the maximum current with only a small voltage drop. Since the generator current is the heaviest, about 9 amperes, a No. 8 gauge wire should be used to carry it. From the distribution cabinets on each floor leads are run to each room; but since usually not more than three lamps are used in any single room, a No. 14 wire is large enough to carry the 3 amperes of current that will be supplied on the room circuits.

The arrangements of the wiring and of the lamps should be made only after careful thought. Lamps should be located where they will be most convenient and efficient. Switches should be placed conveniently, usually near the door through which entrance is made into a room. Three-way switches should be located in halls so that the lights can be turned on and off from either floor; this applies to the basement also.

For years the carbon filament lamp was the only kind of incandescent electric lamp available. Then the tantalum filament lamp was invented, and was quickly followed by the tungsten filament lamp. With the last type of lamp a given amount of energy will produce about three times the candle power that would be furnished by an ordinary carbon filament lamp. A tungsten lamp giving 20 candle-power and using an ampere of current under 25 volt pressure will use a total of 25 watts of energy, or about $1\frac{1}{4}$ watts per candle power; whereas under the same conditions a carbon lamp would consume about $3\frac{1}{2}$ watts. Besides this, the life of tungsten lamps is longer than that of carbon lamps, and the light they give is clearer and more nearly white.

CHAPTER VIII

HEATING FARMHOUSES

THE first essential to comfort in the mind of the average American is ample warmth in all rooms; a cold house is always an uncomfortable house, and, in so far, a cheerless home. The importance of heating and its relation to health has been fully realized only in recent years; if the house is not well heated, all the occupants are uncomfortable, but the children are the ones who suffer the most. Their clothing is of lighter materials, does not well cover the body, and, on account of their activities, more easily becomes disarranged than that of the adult members of the family; they play upon the floor — always the coldest and most drafty portion of the room.

An even temperature indoors, with proper ventilation, and the rational use of heavier clothing to meet a lower temperature without, are the two great essentials of well-being as far as this phase is concerned; the most frequent cause of colds and their attendant ills is *uneven* temperatures and severe drafts within the house. It is not *low* temperature which causes one to "catch cold"; if it were, every one who ventured out in zero weather would become ill. The man who goes outdoors at such a time without adequate protection in the way of clothing is likely to take cold; so is the one who, within his house, changes from a room at 70 degrees to one at 50 degrees, unless at the same time he changes his clothing to compensate for the change in temperature. For this reason the rightly ordered home must be evenly heated.

The aboriginal man, living in caves and rudely constructed huts, found the attainment of an even temperature within his building an impossibility; to be well heated, a house must be well built, and his was not. For the earliest peoples, therefore, the main protection against cold was always clothing, and this is still the main resource of many millions of human beings. But, however great the reliance upon clothing to protect the body against cold, heat from fire has always been an important additional resource.

Two main principles have been followed in the methods of obtaining artificial heat: first, that of maintaining in each room its own individual fire; second, the establishment of a central source of heat, with means for distributing that heat to the various rooms of the dwelling.

The Open Fire

The earliest method of heating was, no doubt, a fire built upon an earthen floor in the middle of the room. An elaboration of this method came with the use of the cresset of the Middle Ages, which is essentially an iron basket designed to confine the fire and raise it above the floor; incidentally this furnished a better draft. This method possessed one advantage which has never been excelled by a heating system, in that all the heat was transmitted to the room; but there was a serious disadvantage in confining the products of combustion in the room itself. It became imperative, therefore, to obtain some relief from this condition, and a hole in the roof of the tent or hut was made, but before the smoke could find the exit, it became more or less distributed within the room. Then a chimney was built for the purpose of taking the smoke directly from the fire and discharging it from the room, the fire burning in an arched opening at the base of the chimney. This was the first form of the fireplace, which has been used in a more effective way ever since in supplying heat and in furnishing an atmosphere of cheer and warmth in a home.

The Fireplace

The old-fashioned fireplace was very large; some were so large as to hold a backlog so heavy that it must be hauled by a yoke of oxen. But these large fireplaces were far from economical, and little by little, especially after the use of coal became more common, they were restricted in size, and the basket grate, which first stood in the center of the wide, deep hearth, was closely arched in, and became the coal grate of modern days. The fireplace has many disadvantages, such as uneven heating, need of almost constant attention, difficulty in handling ashes, danger from fire, and drafty rooms; but the most serious disadvantage lies in its inefficiency as a source of heat, since it constantly forces up the chimney a large amount of heat which does not raise the temperature of the room, and at the same time it steadily draws into the room a large volume of cold air which must be constantly and quickly heated if the temperature of the room is kept up to a comfortable degree.

The construction of an ordinary fireplace is shown in Figure 142. The roof of the fire chamber should not ordinarily be more than 26 inches above the floor, unless it is built especially for burning large logs, when it may be from 30 to 40 inches high, and as wide as necessary. A rough rule by which to gauge the size of a flue is to construct it with the opening one tenth of that of the fireplace opening. If the flue is contracted at the throat of the fireplace, it will insure the thorough heating of the air at this point, and thus greatly improve the draft. By

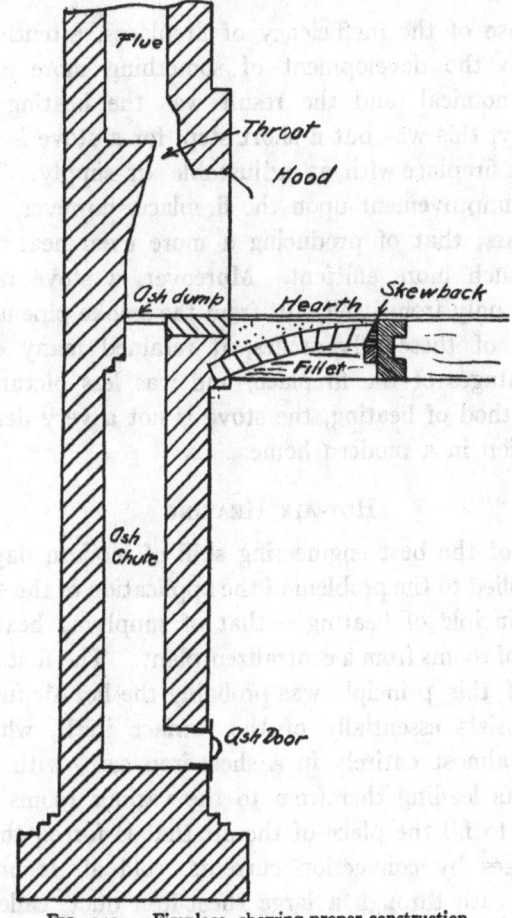

FIG. 142. — Fireplace, showing proper construction.

contracting the throat in this way it is very easy to construct a level shelf in the flue above the fireplace opening; descending currents of air and smoke strike this shelf, rebound, and return up the chimney without puffing out into the room.

STOVES

Because of the inefficiency of fireplaces, attention was given to the development of something more efficient and economical, and the result was the heating stove. In a way, this was but a short step, for a stove is only a portable fireplace with an adjustable air supply. It was a great improvement upon the fireplace, however, in two particulars, that of producing a more even heat and in being much more efficient. Moreover, a stove radiates heat not only from itself, but from the smoke pipe as well. In spite of these advantages, it retained many of the disadvantages of the fireplace, and was less picturesque. As a method of heating, the stove is not a very desirable installation in a modern home.

HOT-AIR HEATING

Some of the best engineering skill of modern days has been applied to the problem of the application of the second great principle of heating — that of supplying heat to a number of rooms from a centralized plant. The first application of this principle was probably the hot-air furnace. This consists essentially of the furnace itself, which is inclosed almost entirely in a sheet-iron case, with sheet-iron ducts leading therefrom to the various rooms to be heated; to fill the place of the air that is forced through these pipes by convection currents, cold air is brought into the case through a large sheet-iron duct, called the cold-air duct.

Hot-air furnaces are all quite similar, differing only in the design and arrangement of the parts; they are all the same in consisting of a steel or cast-iron case, with firebox, grate, and ashpit. Some are fed through a door in the side, the fuel being thrown directly into the firebox; others have special arrangements so that the fuel, which is comparatively small in size, is supplied from below; those possessing this feature are known as underfeed furnaces. The fuel used in hot-air furnaces is almost always coal, either bituminous or anthracite.

Two distinct types of pipes are used for conducting the heated air to the rooms: first, those which are nearly horizontal and lead from the top of the furnace casing — these are usually round and made of a single thickness of bright tin wrapped with two or more thicknesses of asbestos to prevent loss of heat, and are called *leaders;* they should, if possible, be erected with an ascending pitch of one inch to one foot; second, rectangular vertical pipes or risers, termed *stacks*, made in such sizes as will fit in the partitions of buildings and to which the leaders connect. At the bottom of the stack is an enlarged section called the *boot*, which is provided with a collar for connection to the leader. At the top of the stack is a rectangular chamber into which the *register box* is fitted. To lessen fire risk, these boxes should be made with double walls. Each leader should have a damper near the furnace, so that when necessary or desirable it may be closed; the nearer the damper is to the furnace end of the leader, the less will be the danger of superheating.

Provision should be made for evaporating water in the air chamber, to moisten the air forced through the house; most furnaces are equipped with a pan for this very purpose, which is an important one, since warm air requires more moisture than cold to maintain a comfortable degree of saturation. It is a generally accepted but mistaken belief that heat supplied by a hot-air furnace is necessarily a dry heat; all that is necessary is to pass the heated air over water.

The hot-air furnace system of heating possesses certain advantages, principal among which is the readiness with which the temperature can be raised. In cost it is much below that of steam or hot-water heating, two systems with which it is comparable, and it requires no care to prevent bursting of pipes or boiler from freezing. Unless the construction is good, and the erection has been carefully made, combustion gases are likely to be delivered to the rooms, which is, to say the least, annoying; but this objection can be overcome, and cannot be justly considered a disadvantage. The disadvantages of the hot-air system lie in the comparative high cost of operation, in the rapidity with which it loses heat when the fire becomes low, and in the difficulty of even heating on windy days.

The question of the ventilation provided by a hot-air furnace is an important one. The system presupposes a very generous supply of air, which, in properly erected systems, is fresh when brought to the furnaces, is then heated and distributed to the rooms. When so constructed that the air brought to the furnace is taken from the interior of the house itself, the furnace is a source of danger, for the air will become so devitalized that it will be absolutely unfit to breathe. The air that is brought in to the furnace from out of doors is likely to lose a great deal of its supply of oxygen, unless care is taken to prevent the furnace from becoming too hot.

STEAM HEATING

Heating by means of steam came perhaps first as a development of a method of heating to overcome the disadvantages of hot-air heating. The essentials of the system consist of the boiler with the furnace beneath, a system of distributing pipes for the steam, and radiators through which the heat of the steam is liberated into the rooms. The theory concerned in the operation is quite simple: the water in the boiler is heated, and steam is generated which rises through the pipes to the radiators; since it loses heat through the radiators, some condensation will result, and this is either brought back to the boiler or disposed of in some other way.

There are two general systems of heating, in the first of which, known as the Gravity Circulating System, the water of condensation from the radiators flows by its own weight into the boiler at a point below the water line; in the second, the water of condensation does not flow directly back to the boiler, but is returned by special machinery or in some cases wasted. The latter system is sometimes called the High Pressure System, because steam of any pressure can be generated in the boiler, part of which can be used for power purposes. High-pressure steam, however, is seldom used for heating, but is reduced to not more than 10 pounds by throttling from the boiler or by passing through reducing valves; sometimes the exhaust steam from engines and pumps is used.

The boiler for house heating with either steam or hot water should be chosen very carefully. It should be large enough to contain a sufficient amount of water; the firebox should be deep and spacious; it should be easily accessible for cleaning; it should have no joints exposed to the direct action of the fire; a sectional boiler is the better, since no general explosion can occur, should one section give out; the construction should be durable and good, the very best gauges, safety valves, and other fixtures

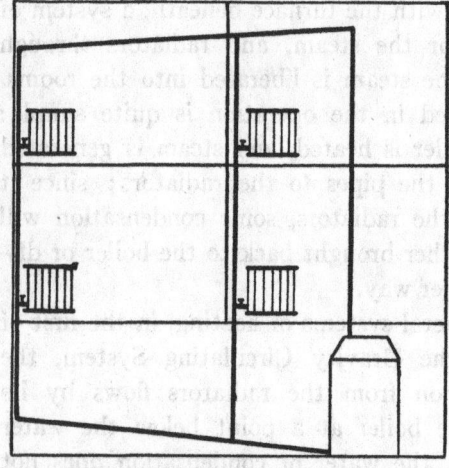

Fig. 143. — Complete circuit system.

should be used, and it should be capable of working to its full capacity with the highest economy.

The systems of piping ordinarily employed provide for either a partial or a complete circulating system, each consisting of main and distributing pipes and returns.

Three systems of piping are in common use.

1. *Complete Circuit System.* — This is sometimes called the "overhead single pipe system," and was first employed in this country by J. H. Mills. In this system the main pipe is led to the highest part of the building, usually the attic, from whence distributing pipes are run to the various return risers, which extend to the basement and discharge into the main return. A

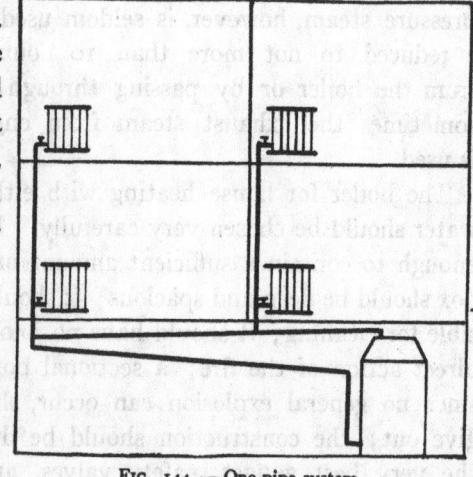

Fig. 144. — One-pipe system.

diagram of this system is shown in Figure 143. The supply for the radiators is all taken from the return risers, and in some cases the entire return circulation passes through the radiators.

2. *Ordinary One-pipe System.* — As shown in Figure 144, in this system a large steam main, elevated close to the ceiling of the basement, runs around to a point where the last radiator is taken off, and is then connected into a return main to the boiler. All the water of condensation returns through the same pipe. This system requires only one connection to each radiator, this being an advantage over the Mills system.

3. *Two-pipe System.* — This system,

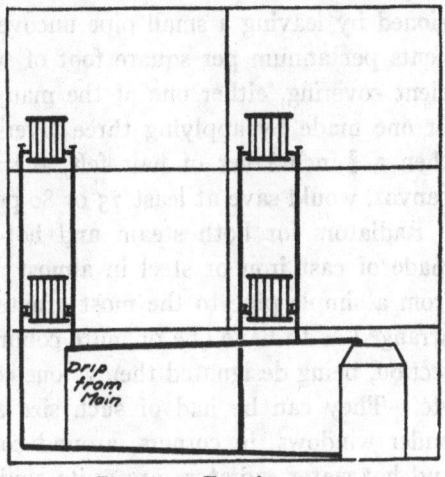

FIG. 145. — Two-pipe system.

shown in Figure 145, consists of steam and return mains in the basement and two connections to each radiator. It is used in large buildings more than in residence heating.

It is difficult to make a definite comparison of the different piping systems, since so much depends upon local conditions. Undoubtedly the complete circuit system gives the freest circulation, and it is applicable either to hot-water or steam heating; it is simple in its construction, and any small error in its installation will not affect its successful operation to any material extent. The fact that the distributing pipes must be placed in the top of the building will in many cases render the system so objectionable that it cannot be used. It would seem that with steam heating only one connection should be necessary for successful operation.

No main steam or hot-water pipe should be left unprotected, for the loss of heat by radiation in such a case is very great. Carpenter estimates the actual loss occasioned by leaving a small pipe uncovered to be about 30 cents per annum per square foot of surface; and an efficient covering, either one of the many commercial types, or one made by applying three layers of asbestos paper, then a $\frac{3}{4}$-inch layer of hair felt, the whole protected by canvas, would save at least 75 or 80 per cent of this.

Radiators for both steam and hot-water systems are made of cast iron or steel in almost any size or variety, from a simple pipe to the most ornate. They may be so arranged as to have one or more columns of water in each section, being designated then as one-column, two-column, etc. They can be had of such size and shape as to fit under windows, in corners, around columns, etc. Steam and hot-water radiators are quite similar in construction, except that the latter have a horizontal passage connecting the sections at the top as well as one at the bottom; this construction is rendered necessary in order to draw off the air which gathers at the top of each section. Hot-water radiators may serve admirably for steam circulation.

Steam-heating plants have been in very successful operation for a number of years, and afford a very good solution of the heating problem. Steam or hot-water plants never cause danger from fire, since no part of the system can become overheated, either through accident or carelessness. The pipe connections are inconspicuous, and sounds and odors cannot be carried through them, as is the case with air ducts. Radiators are sanitary, sightly, noiseless, and can be located in the most convenient place in any room. A steam-heating system is simple and economical in operation, and requires less care than a hot-air system. Since there is only a comparatively small quantity of water in the boiler, it will be only a very short time from the time the fire is built until steam is being generated and circulated through the pipes, thus heating the rooms quickly; but just as soon as the fire dies down, the steam circulation ceases, and the temperature of the rooms falls as rapidly as it had risen. Unless the pipes are carefully installed, *water hammer* is likely to occur; this is caused by water accumulating in low places or pockets in horizontal pipes to such an extent as to condense some of the steam in the pipe, thus forming a vacuum which is filled by a very violent rush of steam and water, causing a severe concussion which sometimes does considerable damage. In the popular mind there is an idea prevalent that steam and hot-water systems necessarily afford moist heat, but such is obviously not the case; in neither of the two systems can moisture get out of the pipes, since they are of course water-tight. Sometimes provision is made for the escape of steam at the valves, but generally this is such an annoyance that the valves are kept closed; so that unless some provision is made for a supply of moisture, steam or hot-water heat will be found to be drier than hot-air heating.

Hot-water Heating

Heating by means of hot water is accomplished by means of circulating hot water in the radiators instead of steam. The principle involved is illustrated in Figure 146. A

U-tube with the legs connected at the top is filled with water, and heat is applied at one side; the heated water is lighter and will tend to rise, crossing over at the connection and occupying the space formerly filled by the cooler water which has now flowed across to fill the space vacated by the heated water; thus a continuous circulation is maintained. Exactly the same phenomenon occurs in the hot-water installation; the entire system, radiators, circulating pipes, and boiler, are filled with water; this water is heated in the boiler. The hot water in the boiler is light, and has a constant tendency to rise, while the water which has lost its heat through the radiators is heavy, and has a corresponding tendency to fall; consequently, a circulation occurs and is maintained as long as the temperature within the boiler is a few degrees higher than that of the house.

FIG. 146. — Hot-water circulation.

Two general systems of hot-water heating are in use; namely, (1) *the open-tank system*, and (2) the *closed-tank*, or *pressure* system. In the former an open expansion tank is connected to the system in such a way as to receive the increase in volume of water due to expansion by heat, and is connected with the outside air by a vent pipe, so that there is no pressure on the tank. In the latter system, a similar tank is used, but the vent pipe is closed, and a safety valve is attached, so that by increasing the pressure on the system, the water may be heated up to the temperature of low-pressure steam, and hence less radiating surface and smaller pipes may be used. With the open expansion tank, about the only chance for an explosion is the stopping of the expansion pipe by freezing or by the closing of a valve in the pipe; and to prevent this, no valve should be placed in the pipe, and it should be well protected from frost. The expansion tank should be located several feet above the highest radiator, and should have a capacity approximately one twentieth of the cubical contents of boiler, pipes, and radiators.

Almost any boiler that can be used for steam heating is suitable for hot-water heating, there being but a slight difference in the interior design to improve the circulation. In an efficient heater the water is separated into small portions so that it may heat quickly, and as little resistance as possible is offered to free circulation. Efficiency in point of fuel consumption is an important feature, as is facility and convenience in cleaning fire surfaces; for a thin coating of soot will materially decrease the efficiency.

Piping systems for hot water are quite similar to those for steam heating, and, as in steam heating, there are three systems in vogue:

(1) The *overhead system*, exactly similar to the Mills system with the exception that two connections are always made to the radiator, one for the inlet and the other for the outlet of the water.

(2) The *two-pipe system*, the one most commonly used, has separate mains and returns.

(3) The *one-pipe system* has a single pipe running around the basement as in the corresponding steam system, except that the main hot-water pipe rises from the boiler; the flow pipes are taken from the top of the main, and the water after passing through the radiators is returned by a separate pipe which is connected with the bottom of the main.

Hot-water apparatus should be kept full of water during the summer months, and only enough supplied during winter to keep it at a safe level. This excludes the air and prevents oxidation or corrosion of the pipes, besides reducing to a minimum the incrustation, which might become serious if allowed to accumulate from several fillings.

Hot-water heating plants are highly satisfactory when properly designed and installed. Hot-water radiators do not become so hot as steam radiators, consequently they do not reduce the humidity to so great an extent. The heat can be kept quite uniform, the system being easily controlled, and any radiators can be shut off without resulting in the snapping or gurgling noises common with steam. The first cost is somewhat higher than of a steam installation, because of the greater radiating surface, larger piping, and more expensive fittings. Unless care is taken when the house is vacant, the water in the system is likely to freeze and seriously damage the plant. On the whole, however, it would appear that for average residences hot-water heating is the most satisfactory.

COMBINATION HOT-AIR AND HOT-WATER HEATING

It is sometimes difficult to heat houses of large size with hot air, especially the rooms distant from the furnace, so some means must be provided to carry the heat to these remote and exposed parts. Thus has been evolved a method of inserting in the combustion chamber of the hot-air furnace a small hot-water heater which will heat the water to be carried by pipes to radiators located in the portions of the house most difficult to heat by warm air. As a rule, where there is any choice, the portions of the house which should be heated by the hot water are the halls, bathroom, and perhaps the rooms on the north or west side of the house.

VACUUM CIRCULATING SYSTEM

In recent years there has been in vogue a system of heating popularly known as "vacuum heating." This is simply a modification of a closed system of steam heating, in which the air is removed and kept from flowing back, thus permitting a circulation above or below atmospheric pressure as desired, the pressure and temperature being dependent upon the amount of fire maintained in the heater. For instance, could the air be removed to such an extent that 26 inches of vacuum be produced, the boiling temperature of the water at this pressure would be only 126 degrees F., and if just sufficient fire were maintained to produce that pressure, the temperature would remain at this point; whereas if more fire were maintained, so as to produce greater quantities of steam, the pressure would rise with a corresponding increase in temperature. Such a system would give all the advantages pertaining to low temperatures and regulation of temperature possessed by hot-water heating, and all the advantages relating to high temperatures, small radiators, and low cost of installation pertaining to the steam system.

DESIGN OF HEATING SYSTEMS

Hot-air Systems. — Apparently there is no reliable rule that can be applied to the design of a hot-air heating system, the rules given by manufacturers varying widely, so to be safe it is best to have the contractor installing the furnace guarantee that the furnace shall heat the building to 70 degrees in zero weather without forcing the furnace. The tables given by different authorities for the sizes of pipes also vary a great deal, and considerable judgment should be exercised in using them.

Carpenter uses a quantity which he designates "equivalent glass surface" in deducing rules for hot-air heating; by this term is meant the area of the glass in the exterior windows and doors of the room plus one fourth the area of the exterior wall surface.

Carpenter's rules are as follows:

1. To find grate area in square inches: Divide equivalent glass surface in square feet by 1.25, or multiply by 0.8.

2. To find area of flue for any room in square inches: Divide equivalent glass surface in square feet by 1.2 for first floor, by 1.5 for second floor, and by 1.8 for third floor.

3. Make area of cold-air duct 0.8 of total area of hot-air flues.

4. Make area of smoke flue in square inches one twelfth of grate area, with one inch added to each dimension.

Steam and Hot-water Heating Systems. — Because of different conditions surrounding the installation of heating apparatus, it is impossible to give any set rule that can be used without modification to satisfy all conditions. It is generally assumed that a pressure of from 2 to 5 pounds will be carried, and a temperature of 180 degrees maintained; when systems are designed for heating with a lower heat temperature at the boiler, as in vacuum heating, it is necessary to provide additional radiation. It is general practice to consider 70 degrees as the standard for inside temperature and zero for the outside; when there is a greater difference between the inside and the outside temperatures, one per cent should be added to the radiation for each degree difference in temperature.

The rule submitted by Carpenter for the proportioning of radiators is as follows:

HEATING FARMHOUSES

To the equivalent glass area add:

$\frac{1}{55}$ of cubic contents for second-floor rooms.
$\frac{2}{55}$ of cubic contents for first-floor rooms.
$\frac{3}{55}$ of cubic contents for large halls.

Multiply this result by:

0.25, if for steam.
0.40, if for hot water.

This will give the radiation required in square feet, and from catalogues in which the radiating surface per section of various types of radiators is given, can be ascertained the number of sections necessary.

Pipe sizes can be determined from the accompanying table:

Size of Pipe	Size of Return	Square Feet of Radiation	
		Steam	Hot Water
1	$\frac{3}{4}$	40	30
$1\frac{1}{4}$	1	100	80
$1\frac{1}{2}$	$1\frac{1}{4}$	150	100
2	$1\frac{1}{2}$	275	200
$2\frac{1}{2}$	2	500	325
3	2	750	500
$3\frac{1}{2}$	$2\frac{1}{2}$	1000	700
4	$2\frac{1}{2}$	1500	1000

Boilers are usually rated for direct cast-iron radiation, in square feet. Most manufacturers are somewhat close in their ratings, so it is advisable to add 25 per cent to the total radiation required in choosing the boiler, so as to provide for emergencies, and to insure an ample supply of heat, even in extremely cold weather, without unduly forcing the boiler.

CHAPTER IX

FARM WATER SUPPLY

PROMINENT among the money and labor-saving devices to which the modern and progressive farmer should give his attention is the individual water system. Strangely enough, while water is the most necessary of all commodities, is used more frequently, in larger amounts and for a greater number of purposes, the old method of carrying water by buckets is so common as to be deplorable, in view of the fact that other arrangements so much more convenient and economical are entirely feasible. In the average home not equipped with a water supply system, not less than fifteen minutes a day must be spent in pumping sufficient water to supply the mere necessities of the household. Fifteen minutes a day in the course of a year will amount to ten days of nine hours each, and the income on ten whole days' efforts in a year will certainly more than warrant the additional expenditure.

All stock thrive better if their water is pure and if they can get plenty of it; so as far as this phase of the matter is concerned, it is a matter of dollars and cents. To the dairy farmer, especially, is water supply important; he uses more water than the general farmer, and must supply it an even temperature all the year round. It is an undisputed fact that the drinking of ice water during winter reduces the vitality of the stock and decreases the amount of milk produced. Hogs, too, are peculiarly susceptible to the dangers of impure water that often are present when the supply of water is insufficient; and in warm weather it is decidedly advantageous to have a carefully cleaned watering place, for the hog will drink a few swallows every twenty minutes if it is within reach. Clean water is equally important in raising healthy poultry; one poultry writer asserts that the water contained in the eggs that are laid annually would fill a canal a mile long, 30 feet wide, and 20 feet deep.

The development of rural water-supply systems has been deplorably slow, considering their importance; the knowledge that good systems are extant and that their principles and operation are satisfactory does not seem to have led to their extensive adoption. Recently there seems to be a stronger tendency toward their more widespread use, probably because of the fact that there has been fostered a definite attempt to improve rural home conditions, and that manufacturers, realizing this, have entered more earnestly into the field of producing really good systems.

Sources of Supply. — In almost all cases the source of rural water supply is either a well or a spring; it is only in rare instances and exceptional cases where circumstances and conditions are especially peculiar that surface water or rain water is used for human consumption. When the source is a spring, it should be protected by a concrete curbing, to prevent the ingress of surface or soil-water that might bring contamination.

Wells are either dug or bored. In the case of a dug well, the diameter must necessarily be great enough to admit of a man working within it, as well as of the necessary hoisting apparatus for removing the earth when any considerable depth is reached. The walls are usually lined with brick or stone masonry, to retain the earth and keep it from entering the well. Wells of this type are comparatively shallow; they are common in regions where the soil-water stands at a high level, and they depend upon the seepage to keep them supplied with water. On account of this circumstance, they are more or less dangerous, since, if the seepage occurs from some stratum which originates at the surface of the ground or at some point near it, there is great likelihood of impurities being carried into the well. Innumerable cases are on record where the cause of a typhoid fever or similar epidemic could be traced directly to some well in which contamination had occurred as a result of transmission of the bacteria through shallow subterranean channels from the vault of an outdoor privy.

Bored wells are the only solution of the water supply problem in regions where no springs exist and the water-bearing strata are so deep that they cannot be reached by digging; a well more than one thousand feet in depth is not at all out of the ordinary. The method of producing a well of this kind is to bore a hole with an augur which will pass through a pipe of the diameter desired. As the hole is bored, the pipe, or "casing," is driven down as fast as the augur removes the earth ahead of it. Special rock drills have to be employed when passing through rock strata, and when an underground bowlder is encountered which deflects the augur, dynamite must be employed to remove it. When a stratum has been reached which bears water in sufficient quantity and of desired quality, the boring is discontinued and a section of pipe called the "screen," closed and pointed at the lower end, and perforated for about three feet of its length, the perforations being protected by a fine brass screen, is inserted within the casing at its lower end so as to penetrate the water-bearing stratum. Water enters through the screen, the meshes of which are fine enough to keep the sand out.

Some sort of a pump, operating through rods which reach to the level that water rises within the well, or deeper, is employed to raise the water to the surface. Bored wells give a supply of water that is almost certain to be cold and pure, since it has passed through sufficient filtering mediums to be thoroughly purified.

Artesian wells are bored wells penetrating water strata of such a nature and conformation that the water as a result of pressure is forced out at the top of the well as in a fountain. The explanation of this can easily be gotten from Figure 147, which shows the water-bearing stratum

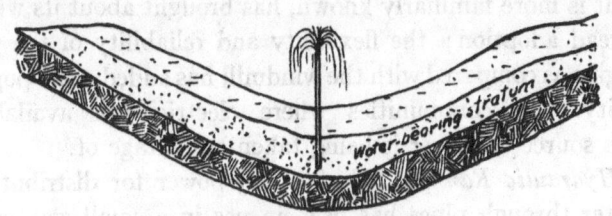

FIG. 147. — Artesian well.

to be in the form of a depression with its ends higher than the top of the well. This type of well is met with in all parts of the country; it derives its name from the fact that investigation was first made of it in the French city of Artois (formerly called Artesium) about 1750.

Cisterns are almost universally used as a storage place for rain water. The method of their construction is similar to that of dug wells, except that in many cases the walls, bottom, and even the top are made of concrete.

Types of Water-supply Systems

The term "water-supply system" may be taken to mean the method by which the water is taken from the source of supply and delivered or distributed at points more or less convenient to the place where it is used. In earlier days, when dug wells were the only type known, the "sweep" and the "wheel and chain" or the windlass, were the only means used to lift the water from the wells. Later, coincident with the development of bored wells, windmills came into use as a power for operating the pumps by which the water was raised. For many years, during the latter part of the nineteenth and through the first decade of the twentieth century, they were practically the only source of power on the farm, but the recent development of the internal combustion engine, or gas engine as it is more familiarly known, has brought about its widespread adoption; the flexibility and reliability of the gas engine as compared with the windmill has added to its popularity. In communities where electricity is available, this source of power is being taken advantage of.

Hydraulic Ram. — This form of power for distributing water through pipes has been in use in a small way ever since its invention by Montgolfier, in 1796, to whom credit is given for having first perfected the automatic machine. Hydraulic rams are in quite common use in localities where conditions are favorable, but they are practically all of small size, designed to raise but small quantities of water, and that to small heights.

Figure 148 is a diagrammatic representation of a simple hydraulic ram. E is the source of supply, A the supply pipe, B the channel, which should be long and straight, a and b the valves, a opening downward and b upward, C the air chest, and D the discharge pipe. Water first flows out in quantity through valve a, but when it has acquired a certain velocity it raises that valve, and the opening is closed. A certain impact results, which raises valve b, and some of the water passes into the air chest C, compressing the air above the mouth of the discharge pipe; the air by its elastic force closes valve *b*, and the water which has entered is raised in the discharge pipe *D*. As soon as the impulsive action is over, and the water in the channel is at rest, the valve *a* falls by its own weight, the flow resumes, and the whole process is repeated.

It will be seen that while a portion of the water is wasted in performing the operation, the power is secured without cost and attention. The water can be raised to a height many times as great as difference in water levels at *E* and *a*; if no energy were lost in friction and in raising the valves, the height of ascent would be to the fall as the quantity which flows out

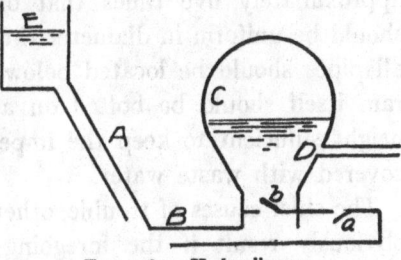

Fig. 148. — Hydraulic ram.

at *a* is to that which is raised, as, for instance, one fifth of the total amount of water flowing out of the channel could be raised to four times the height of the difference in water levels. As a matter of fact, economical operation depends somewhat upon the amount of fall; good practice advises that under ordinary circumstances, one seventh of the water can be raised and discharged at an elevation five times as high as the fall, or one fourteenth can be raised and discharged ten times as high as the fall applied, and so in like proportion as the fall in elevation is varied. One manufacturer gives the following rule for determining the quantity of water which a ram will deliver: multiply the fall in feet by the number of gallons flow, divide this product by twice the height to which the water is elevated, and the result will give the quantity of water (in gallons) which the ram should deliver.

In installing a hydraulic ram there are several precautions which, if observed at the time, will often save trouble later. The upper end of the supply pipe should be a foot or more below the surface of the water, and protected by a strainer or screen to prevent it from becoming clogged. Pipes should be laid straight to reduce friction as much as possible, but if turns are necessary, long bends are better than sharp angles. The length of the drive pipe should be approximately five times that of the vertical fall, and should be uniform in diameter throughout. The ram and all pipes should be located below the frost line, and the ram itself should be bolted on a level foundation, at a height sufficient to keep the impetus valve a from being covered with waste water.

The chief causes of trouble, other than that which would obviously result if the foregoing precautions were disregarded, are imperfect seating of the valves, which can be remedied by grinding, and the filling of the air chest with water. It is essential for the successful operation of the ram that this be prevented, and rams are provided with a small air or "snifting" valve, which admits a certain amount of air at each impulse; if this valve becomes clogged or flooded with waste water, the air chest fills with water and the operation of the ram ceases.

When the supply of usable water is so small that even a small ram would give practically no discharge, and when a more abundant supply of unusable water is available, double-acting or double-supply rams are used. Their operation is identical with that of single-supply rams, the impetus valve being located so that there cannot be in the water discharged any mixed usable and unusable water.

Pressure Systems

The systems that have been described in the preceding pages constitute various methods of transmitting water from the source of supply to the point of consumption or to reservoirs. Modern water supply installations go further than this — they include arrangements for supplying the water under pressure to any part of the farmstead. There are three methods in common use whereby this pressure is obtained, — the gravity system, the hydro-pneumatic system, and the pneumatic system.

Gravity Tank System. — In this system, which is one that has been widely used in the past and which is still employed to a considerable extent, the water is forced into tanks that are elevated higher than the highest water outlet. From these tanks a system of pipes carries the water to all the points at which it is needed. The pressure at the outlet depends, of course, principally upon the height at which the tanks are located.

The gravity system, though of the simplest type, has certain disadvantages. Its value is affected by the weather — in warm weather the water stored in the tanks becomes warm and flat in taste, and in cold weather it is likely to freeze. The tanks themselves are likely to rot if made of wood, and to rust if made of steel, and their supports are often unable to withstand the strain to which they are subjected, and serious accidents sometimes occur when they collapse. Finally, an elevated tank can seldom be given any architectural treatment that will prevent it from being a decidedly obvious and unpleasant feature of the landscape.

Hydro-pneumatic System. — This system was evolved to overcome the inherent disadvantages of the gravity tank system. The essentials of the system, as shown in Figure 149, comprise a hydro-pneumatic pump, a storage tank, and a distributing system of pipes. The pump is so constructed that it can be arranged to pump either air or water into the storage tank; the tank itself is built of sheet steel, and will usually carry a sustained pressure of 150 or 175 pounds without any difficulty.

In the operation of this system, water is pumped into the tank, compressing the air with which the tank is filled;

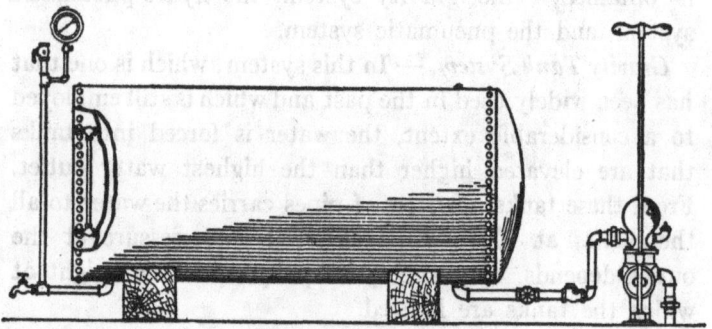

Fig. 149. — Hydro-pneumatic system.

the air is thus constantly exerting a pressure upon the water, and when any faucet in the pipe system leading from the tank is opened, the water is forced out by the pressure of the air within the tank. The proper pressure which should be maintained in the tank can be easily calculated, if we know the height of the highest outlet. Since the tank is usually located in the basement, the highest outlet is not usually more than 25 feet above the tank. The air within the tank before water is pumped in is at a pressure of one atmosphere, or about 15 pounds; this is the equivalent to the pressure of a column of water 34 feet high; a column of water 25 feet high will exert a pressure slightly in excess of 10 pounds. Then to force the last drop of water out against atmospheric pressure at the highest outlet, the pressure within the tank at the moment of emptying should be 15 plus 10, or 25, pounds. This, then, should be the pressure within the tank before any water is pumped into it, if all that is pumped in is to be driven out at a pressure not less than 25 pounds. As the tank is filled, the air within is gradually compressed until it reaches a pressure of from 75 to 100 pounds, the pressure usually maintained in tanks of this type.

To calculate the size of tank necessary for any installation, we make use of Boyle's law regarding the elasticity of gases; *i.e.* the temperature remaining the same, the volume of a gas varies inversely as the pressure; or, pressure times volume is a constant. Expressed as a formula, it is

$$pv = k, \quad \text{or} \quad p_1 v_1 = p_2 v_2.$$

Let us assume that we desire to install a tank which will hold 240 gallons of water at 60 pounds (gauge) pressure. The initial pressure, as noted above, is 25 pounds; the final pressure is 60 pounds plus the pressure of one atmosphere, or 75 pounds; the initial volume is unknown, as is the final volume, but we can express the final volume in terms of the initial volume as $v_1 - 240$, and thus have only one unknown quantity in the equation.

Then
$$25 v_1 = 75 (v_1 - 240),$$
$$25 v_1 = 75 v_1 - 18000$$
$$50 v_1 = 18000$$
$$v_1 = 3600.$$

That is, the total capacity of the tank to fulfill the required conditions is 360 gallons.

The hydro-pneumatic system has proved to be a very popular and successful system and, as developed by several manufacturers, leaves little to be desired in the way of good operation.

There is a slight leakage of the air imprisoned within the tank, since some of it is carried away by the outgoing water; for this reason it is occasionally necessary to pump in a small quantity of extra air, to prevent the pressure from falling below the desired standard. The chief objection to the system is that there is likely to be an accumulation of sediment in the bottom of the tank; this objection is not very great, however, for even should there be any accumulation, it can be easily removed through special openings for the purpose. The great advantages of the system lie in the fact that it can be installed within the basement of a residence, thus avoiding danger of freezing, and that any reasonable degree of pressure can be maintained, at a requirement of only a few minutes' attention each day, or even less often, depending upon the size of the tank.

Pneumatic System. — The pneumatic system derives its name from the fact that while there is a storage tank as an essential part of the system, it contains nothing but compressed air. The other essentials of the system, besides the air-storage tank, are an air compressor, a specially designed automatic pump, and a system of distributing pipes leading therefrom. Figure 150 is used to illustrate the operation of the pneumatic system. The submerged pneumatic pump is in reality a double cylinder contrivance, with connecting valves, each cylinder having air-supply and exhaust pipes and a water-discharge pipe. While one cylinder is filling, the exhaust pipe being open and the air-supply pipe being closed, the other one is discharging under pressure the water contained within itself, the exhaust pipe being closed and the air-supply pipe from the pressure tank being open. As soon as the second cylinder is empty, the valves are automatically operated, the exhaust opening and the air supply closing; the oppo-

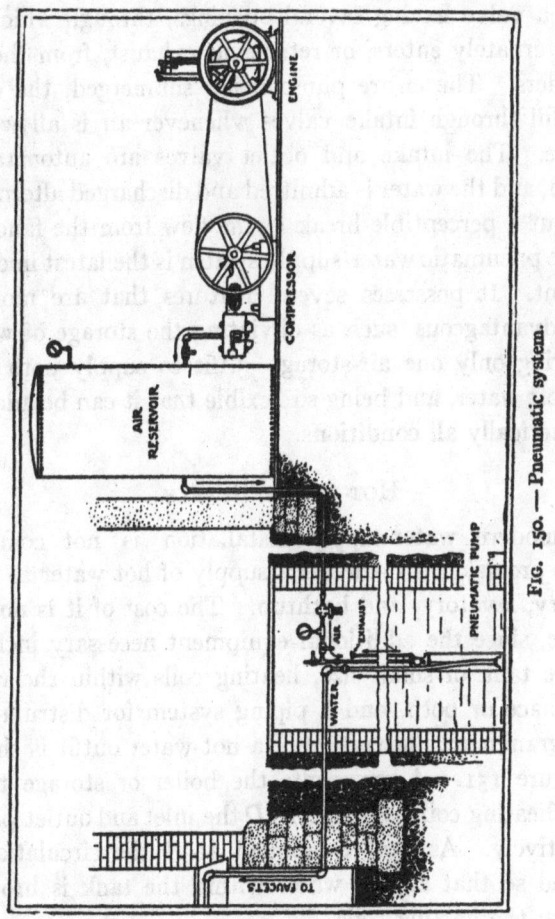

FIG. 150.— Pneumatic system.

site occurs in the first cylinder, the exhaust closing and the compressed air now driving out the water contained therein through the discharge pipe.

All the necessary opening and closing of valves is done automatically by the pressure of the air itself. The automatic mechanism is placed above the cylinders, and contains a valve having several openings, through which the air alternately enters, or returns as exhaust, from the two cylinders. The entire pump being submerged, the cylinders fill through intake valves whenever air is allowed to escape. The intake and outlet valves are automatic in action, and the water is admitted and discharged alternately without a perceptible break in the flow from the faucet.

The pneumatic water-supply system is the latest in development. It possesses several features that are more or less advantageous, such as obviating the storage of water, requiring only one air-storage outfit to supply both hard and soft water, and being so flexible that it can be adapted to practically all conditions.

Hot-water Supply

A modern water-supply installation is not complete unless provision is made for a supply of hot water to sink, laundry, lavatory, and bathtub. The cost of it is not excessive, since the additional equipment necessary includes storage tank of small size, heating coils within the range or furnace or both, and a piping system for distribution. A diagrammatic illustration of a hot-water outfit is shown in Figure 151. A represents the boiler or storage tank, B the heating coils, and C and D the inlet and outlet pipes, respectively. As heat is applied at B the circulation is induced so that all the water within the tank is brought through the heating coils.

Ordinarily the water stored in range boilers, for use in the home, is heated in a water back located in the fire box of the kitchen range, or in coils of pipe in the furnace. The latter plan, of course, results in heating the water only at that season of the year when the furnace is in use, so it is advisable to have a water back in the kitchen range as well, so that advantage can be taken of the daily cooking fires. A water back is simply a hollow casting with two tapped openings for the inlet and outlet pipe connections, as shown

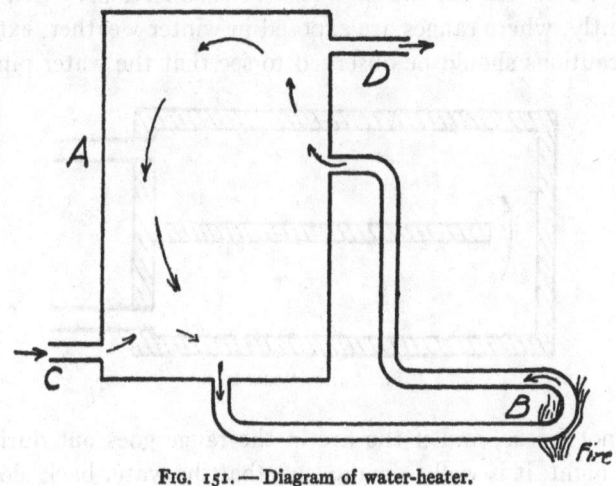

FIG. 151.— Diagram of water-heater.

in Figure 152. Sometimes it has a partition cast horizontally part way across so that the water will be given a more extended circulation. The opening for the outlet, or return pipe, should be made close to the top wall of the water back so that the very hottest water can flow out and not be trapped to become converted into steam. Water backs are made much thicker than would seem necessary to withstand the pressure to which they are subjected, but this pressure can never be determined, as it may vary from a pressure never above 20 pounds per square inch to a street main pressure of 100 pounds per square inch, or more.

Then, too, the casting is subjected to severe shrinkage strains at times and to the strains resulting from the cold water within and the intensely hot fire without. To provide for all this, they are ordinarily designed to withstand an ultimate pressure of 700 pounds per square inch. The most common cause of water backs bursting is freezing of the water in the water backs or connections. Consequently, where ranges are exposed in winter weather, extra precautions should be observed to see that the water pipes

FIG. 152. — Water back.

do not freeze, and if the fire in the range goes out during the night, it is well to make sure that the water back, flow, and return pipes are free from ice before firing is started in the morning.

Where gas is available, a contrivance known as the automatic instantaneous water heater renders the use of water backs and pipe coils unnecessary. This consists of several coils of thin copper pipe, occupying a space inside a casing and placed immediately over a set of Bunsen burners; by this construction, almost all of the heat developed by the combustion of the gas is absorbed by the coils and transmitted to the water. The arrangement also includes a combination automatic gas and water cock, and a thermostat to shut off the supply of gas, but still leave the water flow, when the temperature of the water exceeds a certain degree.

The operation of the heater is quite simple. When any hot-water faucet in the system is opened, the pressure in the automatic valve is relieved, and a supply of gas is admitted to the Bunsen burners, which is ignited by the small pilot light which is always burning in close proximity; in a few seconds the water within the coils is heated. The water-controlled gas and water valve is unique in that the flow of gas and of water are adjusted to each other in such a way that the flow of water through the copper coils is proportional to the amount of gas being consumed, so that just sufficient gas is consumed to heat the water required. When the water has reached the desired temperature, or the flow of water has ceased, the gas is automatically shut off.

As ordinarily constructed, automatic instantaneous heaters are rated to heat one gallon of water for each cubic foot of gas consumed, from ordinary temperature to 130 degrees F., which is about the right temperature for domestic water supply.

CHAPTER X

PLUMBING AND SEWAGE DISPOSAL

PLUMBING systems for buildings include not only the drainage systems, whose purpose is to remove wastes, but the water-supply systems described in the previous chapter as well. Bacteriological investigations show that more disease results from the bacteria carried in through the water supply than from the drainage system, so every precaution must be taken in the installation of the former to make it as safe as possible. However, improperly constructed drainage systems are a source of great danger, and much can be done to increase their efficiency and safety by selecting good equipment and having it correctly installed.

Drainage systems include the house sewer, house drain, soil waste and vent stacks, fixtures, and fixture connections. The house sewer, generally made of tile pipe or cast-iron pipe, extends from the street sewer to a point not less than five feet from the outside of the foundation wall where it connects with the house drain; it also receives the discharge from roof gutters, foundation and area drains, and cellar drains. The house drain, constructed of iron pipes which should be given an asphalt coating both inside and out, is a system of horizontal piping inside the cellar or basement of a building, that extends to and connects with the house sewer; it receives the discharge of sewage from all soil and waste lines, and sometimes rain water from roof gutters. Every house drain should have a main drain trap located just inside the foundation wall, and should have a clean-out ferrule at the end that turns up to connect with a soil or waste pipe. A house drain should never be less than three inches in diameter.

Soil stacks are those that receive the discharge from water closets and urinals, although they may also receive the discharge from other fixtures; they connect with the house drain at the lower end, and their upper end extends above the roof; soil pipes are the connecting pipes between closets or urinals with soil stacks. Waste stacks are similar to soil stacks, except that they are connected with fixtures other than closets or urinals; the connections are called waste pipes. A vent stack is a special ventilating pipe extending from a point below the lowest fixture up to a point above the highest fixture, while it may connect with the stack or extend separately through the roof; its purpose is to supply air to the outlet of fixture traps, thus preventing the water seal being broken by siphonage or back pressure; the connecting pipes between it and the traps are called vent pipes.

There are two systems of stacks and branches in use at the present time; namely, the *one-pipe system* and the *two-pipe system*. In the former a single stack is employed, to which all fixtures are connected, using non-siphon traps, or traps in which the seal cannot be broken by siphonage. This system is much more economical than the two-pipe system, since the cost of the roughing, or putting in the stacks and pipes, is only about half of that of the latter, and the cutting of walls and floors is minimized. From a sanitary standpoint it is satisfactory, and the only objection to it is the possibility of a slight gurgling noise in the waste pipes when a fixture is flushed. In the two-pipe system siphon traps are used, and a vent stack is provided by which the seals of the traps are protected by a system of vent pipes; the more direct the pipes are run and the fewer fittings used, the more satisfactory will be the operation.

In ordinary residences the diameter of soil pipes and stacks is almost always three inches, because water closets are now made with traps not more than three inches in diameter, and because the stacks can be easily concealed in the walls of the building. Soil pipes and waste pipes should always be the full size of the waste outlets from the fixtures, the outlets being large enough to permit the fixtures being emptied quickly so as to thoroughly flush the pipes. The following table gives the sizes of soil or waste pipes and of the corresponding vent pipe for common fixtures:

Fixture	Diameter in Inches	
	Soil or Waste Pipe	Vent Pipe
Water closet	3	2
Bathtub	$1\frac{1}{2}$	$1\frac{1}{2}$
Lavatory	$1\frac{1}{2}$	$1\frac{1}{2}$
Bidet	$1\frac{1}{2}$	$1\frac{1}{2}$
Shower bath	$1\frac{1}{2}$	$1\frac{1}{2}$
Sitz bath	$1\frac{1}{2}$	$1\frac{1}{2}$
Kitchen sink	$1\frac{1}{2}$ to 2	$1\frac{1}{2}$
Slop sink	2 to 3	$1\frac{1}{2}$ to 2
Laundry tub	$1\frac{1}{2}$	$1\frac{1}{2}$
Urinal	$1\frac{1}{2}$	$1\frac{1}{2}$
Drinking fountain	$1\frac{1}{4}$	$1\frac{1}{4}$

A trap is a contrivance which is used to prevent the passage of air or gas through a pipe without materially affecting the flow of sewage. To give good results and to perform their function properly, they should be either made so they cannot be siphoned or have their seals broken by back pressure; they should be sufficiently deep to withstand a long period of loss by evaporation without breaking the seal; and they should be self-scouring so that no deposition of solid sewage can occur. Figure 153 illustrates some common types of traps.

The selection of the fixtures used in a plumbing installation is of considerable importance. To be perfectly sanitary, they must be of some non-absorbent, non-cor-

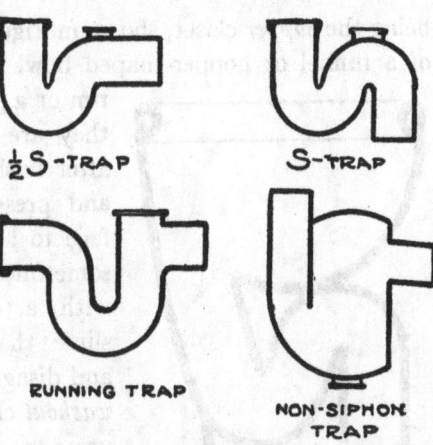

FIG. 153.—Types of traps.

rosive material that will not easily craze, crack, or break, and that has smooth surfaces to which soil will not adhere so firmly that it cannot be broken. Strainers or crossbars should obstruct outlets as little as possible, so that a scouring flow will not be prevented. Overflow outlets should be provided for those fixtures whose waste outlets have stoppers. The plumbing for all fixtures should be free and open; that is, not hidden by woodwork or other casings that would cut off light and air.

Water Closets. — Water closets are made either of solid porcelain or of porcelain-enameled iron, the former being preferable, since they will neither stain nor get foul and can be obtained at the same price, or even less than the latter. They are made in various forms, the simplest

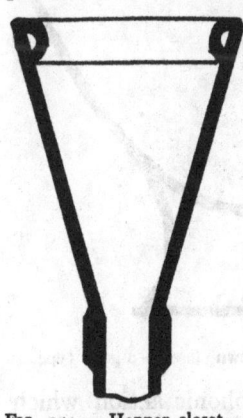

FIG. 154. — Hopper closet — an undesirable type.

being the *hopper* closet, shown in Figure 154, which consists of a funnel or hopper-shaped bowl fitted with a flushing rim or a pipe-wash connection; they are not desirable, since after flushing they are left dry and present a maximum surface to be soiled; this surface sometimes becomes covered with a coating of bacterial slime that is extremely foul and disagreeable in odor. The *washout* closet is shown in Figure 155; it is superior to the hopper closet, but the body of water retained in this type of closet is so shallow that fecal matter is not submerged, consequently offensive odors are given off. The *washdown* closet, Figure 156, is a good type when properly designed so as to give a sufficient depth of water to prevent odors and to prevent any interior surfaces from becoming soiled. The best type of closet yet designed is the siphon-jet closet shown in Figure 157, in which some of the water flows through a jet at the bottom of the seal, starting siphonic action which empties the bowl quickly and completely.

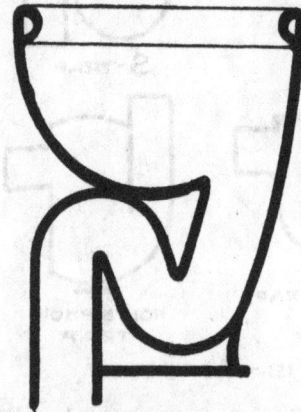

FIG. 155. — Washout closet — note shallow receptacle.

FIG. 156. — Washdown closet — a good type.

Water closets should be so constructed that no woodwork surrounds them; they are usually connected to the soil pipe by means of a cemented joint. The seats should be about an inch thick, of hard wood finished with several coats of good spar varnish. Soft-wood seats with white enamel paint are not satisfactory, since they are easily discolored by urine and gases about the closet. Flush tanks may be located either near the ceiling or just back of the closet, the latter form being preferable on account of accessibility for repair; the flush connection in the latter form is larger than in the high tanks, to compensate for the loss in head.

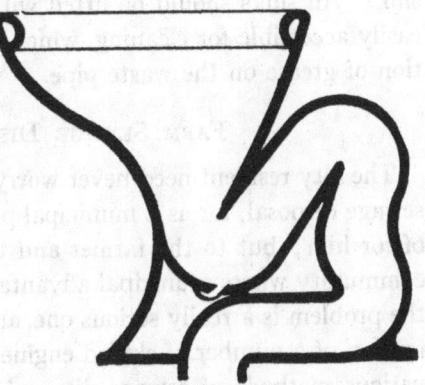

FIG. 157. — Siphon-jet closet — the best type.

Bathtubs and Lavatories. — These should have a smooth, impervious surface, large unobstructed outlet, an overflow channel accessible for cleaning, and no crevices for the accumulation and retention of dirt. Bathtubs are usually made of porcelain-enameled iron, since their construction is simple and subjected to no severe usage, and their cost would be too high for ordinary installations were solid porcelain used in their construction. Lavatories are better made of solid porcelain, but cheaper ones giving very satisfactory service can be obtained in enameled iron construction. The same applies to sitz baths and bidets.

Sinks. — The best sinks are made of porcelain-enameled iron and of solid porcelain, in almost any desired size.

Kitchen sinks, when used for dish washing, should have a draining tray attached, and should have a rubber mat or wooden grating on the bottom so they will be less destructive to china or glassware that is accidentally dropped in the sink. All sinks should be fitted with a grease trap that is easily accessible for cleaning, which will prevent accumulation of grease on the waste pipe.

Farm Sewage Disposal

The city resident need never worry about the problem of sewage disposal, for as a municipal problem it is taken care of for him; but to the farmer and the resident in a small community where municipal advantages are not to be had, the problem is a really serious one, and has occupied the attention of a number of skilled engineers, who have evolved various methods of sewage disposal which experience has shown to be more or less successful in operation.

One of the most undesirable, and certainly the most disgusting and insanitary, features of perhaps 95 per cent of the farms in this country is the privy as it is ordinarily found, bare, unprotected, a breeding place for flies, and a source of danger from all kinds of transmissible diseases. Much has been written about the sanitary privy, and many have been the schemes for devising one, but the best is only a makeshift, and possesses many of the inherently bad defects of all privies.

The installation of a water-supply and plumbing system renders necessary the provision of some method of taking care of the wastes from the various fixtures. In some localities cesspools are used, endangering the water in surrounding private wells and creating an unwholesome condition of the subsoil in the immediate vicinity of the buildings. Such a condition is inexcusable, yet is often permitted to continue for years, and as soon as one cesspool refuses longer to do its work, another is dug, until the ground is often honeycombed with these pits. Sometimes the sewage is discharged into storm drains, with the result that stoppages are frequent and the subsoil is seriously contaminated by leakages through imperfect joints. In other localities is followed the very dangerous procedure of discharging sewage into streams which in dry weather may be but a trickling stream; this practice is objectionable even when the stream contains sufficient water to effect a considerable dilution of the sewage.

Cesspools afford a popular and simple method of getting rid of sewage, but there is a very prevalent misunderstanding as to their safety and effectiveness. As a matter of fact, they may apparently completely dispose of the sewage, but they cannot be considered sanitary, since in time the soil surrounding them will become what is known as "water-logged," and will retain decomposed sewage to such an extent as to be extremely objectionable. There is always the possibility of the unpurified and possibly infected sewage finding its way into and contaminating an underground body of water from which wells, even at a considerable distance, derive their supply.

Acceptable methods of disposing of sewage in a sanitary manner are as follows:

1. By irrigation.
2. By the use of some form of a septic tank.

Irrigation, as applied to sewage disposal, may be either of the surface or subsurface type. Surface irrigation, as the name implies, consists of the discharge of the sewage upon the surface of the ground, necessarily at some considerable distance from occupied buildings and from any well or source of water supply. To prevent saturation of the subsoil, it is usually necessary to place underdrains of common drain tile beneath the irrigation beds, at a depth of from four to six feet below the surface of the ground, and with a good outlet into a ditch or some other water course. Subsurface irrigation does not differ much from irrigation proper, save that the sewage is applied beneath the surface of the ground. For the proper operation of either of these systems the land to be irrigated should be completely dry and porous in character, and in size large enough to provide for two or more beds of equal size, so that sewage may be diverted from one bed to another at frequent intervals, to allow the land last irrigated to rest and regain its normal condition.

The *septic tank* method of sewage disposal is, in fact, the most scientific, perfect, and efficient system of sewage disposal yet devised. Until recently it was confined almost exclusively to the disposal of the sewage of cities and large villages, and of buildings and institutions of a public character. Though its application to small plants is as yet in a more or less experimental stage, still the developments that are available are much superior to any other methods yet devised.

The modern theory of complete sewage disposal is exemplified in these small plants. Sewage, in general, is a complex liquid containing organic as well as inorganic matter both in solution and suspension; the sewage from households consists of kitchen, laundry, and bathroom wastes, dirty water from scrubbing, and human wastes from closets. Modern biological methods of sewage purification are based upon the fact that all sewage contains numberless bacteria, most of which are not only harmless, but useful in acting upon sewage material. These bacteria are of two classes, the *anaërobic* bacteria which live and multiply only in the absence of light and air, and the *aërobic* bacteria, which require the oxygen of the air to live and functionate. The anaërobic bacteria liquefy and gasify the organic matter in suspension in the sewage, while the aërobic bacteria act upon the organic matter in solution and assist in the processes of oxidation and nitrification.

The sewage treatment comprises two stages:

1. A preliminary process for the removal of the organic matter in suspension; this necessitates as part of the

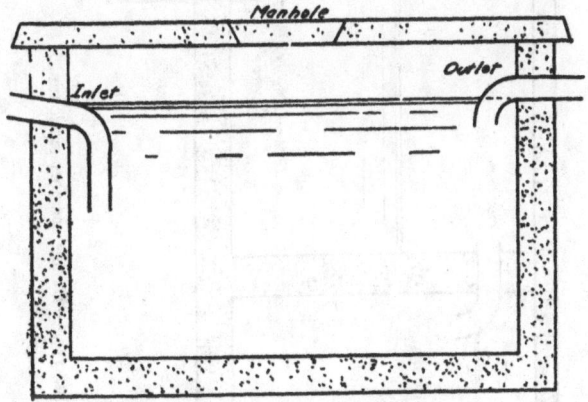

FIG. 158. — Single chamber septic tank.

apparatus a chamber known as the "septic tank," in which the anaërobic bacteria are given an opportunity to perform their functions.

2. A purification process whereby the oxidation and nitrification of organic matter in solution are accomplished; this is carried on in filter beds or in subsurface irrigation systems where sufficient air is available for the successful operation of the aërobic bacteria.

Septic tanks consisting of only one compartment, as shown in Figure 158, have given excellent results in

340   FARM STRUCTURES

actual use. As long as the single compartment tank is not disturbed, the bacterial processes occurring in it are not disturbed, and its operation is successful. Within a

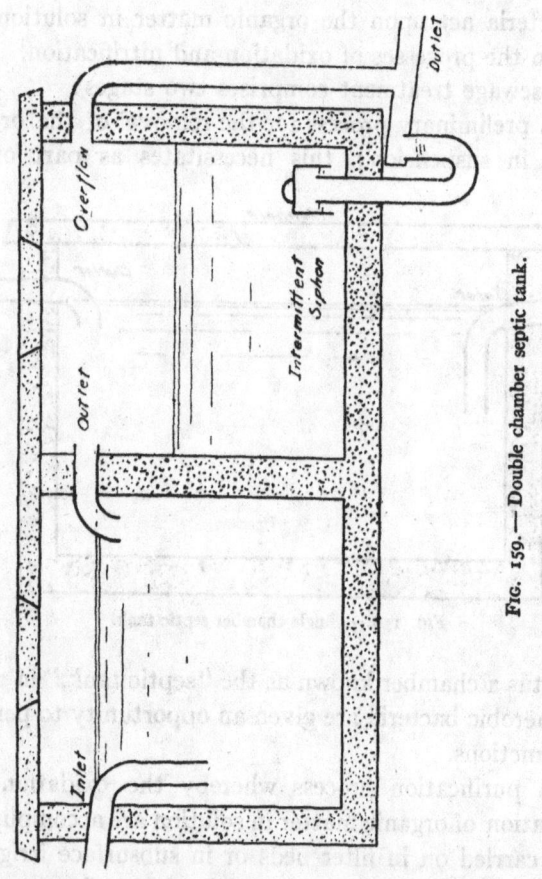

Fig. 159.—Double chamber septic tank.

short time after sewage enters the tank, a scum will form on the surface an inch or more in thickness, consisting of a solid mass of putrefactive bacteria, and which serves to keep out the air so that the anaërobic bacteria can work. Thus scum should not be disturbed, for disturbance will retard the putrefaction process. For this reason a single chamber tank is not entirely satisfactory, because whenever it is desired to clean the tank, the scum must be disturbed. A double chamber tank, as illustrated in Figure 159, is more desirable, since it will provide an undisturbed settling chamber, besides permitting of the installation of an intermittent flow siphon, which is a decidedly advantageous feature.

The size of the tank can be built to suit conditions; 25 gallons of sewage per person per day will give a basis upon which the size of the tank can be proportioned. A width of 3 or 4 feet and a depth the same is satisfactory, and the length can be made to depend upon the amount of sewage to be taken care of. Usually it is well to have the tank large enough so that the sewage may remain in it for two or three days, thus insuring a bacterial action sufficient to liquefy the sewage almost completely. The tank should be built of some water-tight material, preferably concrete, to which has been added some form of commercial waterproofing.

From the tank the liquefied sewage is discharged into a subsurface drainage system, where the aërobic bacteria are enabled to complete the disposal process. This system consists simply of ordinary drain tile, laid in the ground not deeper than a foot or 16 inches from the surface of the soil to the top of the tile, with loose, protected joints. The length of the drainage system will depend upon the character of the soil, stiff, clayey soils requiring that 5 feet of tile be laid for each gallon of sewage discharged, while open, porous soils will require only 3 feet of tile for the same amount of sewage.

If the tank were so constructed that there might be a continual seepage of sewage from it, as would be the case with the tank shown in Figure 152, the soil surrounding the drainage system might become so water-logged that no oxygen could penetrate it, thus destroying the action of the aërobic bacteria. To obviate this difficulty, a device known as the automatic intermittent flow siphon is installed in a second chamber; this siphon can be so adjusted that it will operate only when the depth of the sewage in the siphon chamber is sufficiently deep to start siphonic action. Since the siphon is adjustable, it can be made to discharge as often as desired, permitting, in the meantime, the soil surrounding the drain tile to dry out and absorb a new supply of fresh air. The objection might be made that in cold climates the sewage in the distributing tile might freeze, thus preventing the operation of the system; but experience has shown this objection to be unfounded, the gases arising from the sewage generating sufficient heat to counteract cold and prevent freezing.

INDEX

Acetylene, 288.
  burners, 289.
  generator, 289.
  manufacture, 288.
Air-gas lamps, 287.
Architectural styles, 260.
  American, 263.
  California, 263.
  Colonial, 261.
  Dutch, 262.
  English, 262.
  mission, 263.
Artesian wells, 317.
Ashlar masonry, 18.

Balloon framing, 86.
Bark, 2.
Base, for paints, 55.
Bastard sawing, 9.
Bathroom, 269.
Bathtub, 335.
Beamed ceiling, 110.
Bedroom, 268.
Belted, 7.
Bent, 221.
Bessemer process, 15.
Blocks,
  concrete, 46.
    curing, 47.
    design, 47.
    laying, 48.
    types, 46.
  vitrified clay, 151.
Board measure, 10.
Bond, 27.
Boot, 301.
Box sill, 85.
Braced framing, 87.
Brash, 7.
Brick, 23.
  classification, 24.
  measurement, 25.
  size, 25.
  strength, 25.
  weight, 25.

Bridging, 86.
Broken ashlar, 19.
Broken stone, 33.
Building construction, 79.
Building location, 67.
Building materials, 1.
Bungalow, 263.
Burners, acetylene, 289.
Byrkit lath, 93.

California style architecture, 263.
Candles, 285.
Casement windows, 108.
Cement, 32.
Colonial architecture, 261.
Commercial laying house, 199.
Complete circuit system, 304.
Concrete, 31.
  blocks, 46.
  block silo, 166.
  coloring, 40.
  definition, 31.
  finish, 39.
  floors, 124, 161, 188, 204, 239, 254
  forms, 38, 82, 162.
  foundation, 81, 159.
  materials, 32.
  mixing, 34.
  properties, 36.
  proportioning, 35.
  reenforced, 48.
  silos, 156.
  strength, 51.
  waterproofing, 37.
Cornice, 100.
Crown glass, 63.
Curtain front, 189.

Defects of wood, 5.
  belted, 7.
  brash, 7.
  dry rot, 5.
  heart-shake, 6.
  knotty, 7.
  rindgall, 7.

INDEX

Defects of wood (*continued*).
  star-shake, 6.
  twisted, 7.
  upset, 7.
  wet rot, 5.
  wind-shake, 6.
  worms, 5.
Design of joists and girders, 226.
Dietrich's swine house, 208.
Dining room, 266.
Doors,
  residence, 106.
  silo, 176, 156, 146.
Dormers, 101.
Drains, 330.
Drier, 57.
Dutch architecture, 262.

Efflorescence, 28.
Electric lighting, 291.
  design of system, 292.
  equipment, 292.
  lamps, 295.
Enamel paint, 59.
Endogenous stem, 1.
English architecture, 262.
English bond, 27.
Estimating, 112.
Excavating, 80, 112.
Exogenous stem, 1.

Farm building ventilation, 273.
Feed rack for sheep, 215.
Fireplace, 299.
Flemish bond, 27.
Floor deafening, 98.
Floors,
  barn, 226, 239, 254.
  granary, 119.
  machine shed, 125.
  poultry house, 187.
  residence, 96, 114.
  sheep barn, 214.
  silo, 161.
  swine house, 205.
Flues,
  hot-air, 301.
  ventilating, 279, 282.
Foundations, 79.
Framing, general, 84.
  balloon, 86.
  barn, 219.
  braced, 87.

  floor opening, 88.
  poultry-house, 192.
  roof, 90.
  round barn, 233.
  silo roof, 178.
  swine house, 208, 211.
Fuller's rule, 36.

Gable roof, 90.
Gambrel roof, 222.
Gas heater for water, 328.
General purpose barn, 256.
Glass, 61.
  crown, 63.
  ground, 63.
  plate, 62.
  prismatic, 63.
  sheet, 62.
Granaries, 118.
  arrangement, 120.
  equipment, 120.
  floors, 119, 120.
Gravel, 33.
Gravel roofing, 30.
Gravity tank, 321.
Gurler silo, 149.
Gutters,
  barn-floor, 238.
  roof, 100.

Heart-shake, 6.
Heartwood, 2.
Heating systems, 296.
  combined hot-air and hot-water, 310.
  design, 311.
  essentials, 296.
  fireplace, 298.
  hot-air, 300.
  hot-water, 307.
  open-fire, 297.
  steam, 302.
  vacuum, 311.
Hollow walls, 28.
Hopper closet, 333.
Horse barns, 252.
  design, 252.
  essentials, 252.
  floors, 254.
  measurements of stalls, 253.
  small, 256.
  stallion, 256.
  ventilation, 254.
Hot-air heating system, 300.

INDEX

Hot-water heating system, 307.
Hot-water supply, 326.
Hydraulic ram, 318.

Ice houses, 132.
　construction, 134.
　concrete, 135.
　inexpensive, 134.
　types, 133.
Ice storage, 133.
Ideal dairy barn, 245.
Indirect radiator, 278.
Individual swine house, 206, 207.

Joints,
　brick masonry, 26.
　stone masonry, 19.
Joists, 85.
　design, 226.
　hangers, 89.

Kerosene lamps, 286.
Kiln drying, 8.
King system of ventilation, 280.
　design, 282.
　flues, 282.
　principles, 281.
Kitchen, 264.
Knotty wood, 7.

Laminæ in stone, 18.
Lamps,
　acetylene, 289.
　air-gas, 287.
　electric, 295.
　kerosene, 286.
Lath, 93.
Lavatories, 335.
Leader, 301.
Lighting farm buildings, 285.
　acetylene, 288.
　air-gas, 287.
　candles, 285.
　electricity, 291.
　kerosene, 286.
Living room, 267
Location of farm buildings, 67.
　advantage of good, 76
　application of principles, 73.
　principles, 69.

Machine sheds, 123.
　arrangement, 126.
　floors, 124.
　roof framing, 125.
Mission architecture, 263.
Mixtures for concrete, 35.
Mortar, 52.

Nails, 63.
　classification, 63.
　cut, 63.
　holding power, 64.
　sizes, 66.
　special, 63.
　wire, 63.
　wrought, 63.

One-pipe system, 305, 309.

Paint, 55.
　application, 58.
　composition, 55, 57.
　definition, 55.
　enamel, 59.
Painting, 58.
Pigment, 57.
Pipe sizes for plumbing, 332.
Pitch, 91.
Pit silos, 185.
Plate, 89.
Plinth, 107.
Plumbing systems, 330.
Pneumatic water-supply system, 324.
Portable colony house, 198.
Poultry houses, 186.
　colony type, 198.
　convenience, 190.
　commercial, 199.
　feed boxes, 191.
　floor, 188.
　for average farm, 194.
　foundations, 187.
　general construction, 192.
　location, 186.
　roofs, 193.
　sunlight, 190.
　types, 193.
Prismatic glass, 63.

Quarter sawing, 9.

Rafter, 89, 90.
Ready roofing, 31.
Reënforced concrete, 48.

Residence, 257
  architectural styles, 260.
    American, 263.
    California, 263.
    Colonial, 261.
    Dutch, 262.
    English, 262.
    mission, 263
  bathrooms, 269
  bedrooms, 268.
  dining room, 266.
  kitchen, 264
  living room, 267.
  typical plan, 269
  ventilation, 274.
Rifts in stone, 18.
Rindgall, 7.
Roof framing, 90.
Roofing, 29.
  gravel, 30.
  ready, 31.
  shingle, 29.
  slate, 29.
  tile, 30.
  tin, 30.
Roofs,
  barn, 222.
  gable, 223.
  gambrel, 223.
  poultry house, 193
  residence, 90.
  silo, 176.
  swine house, 208.
Roosts, 191.
Round barn, 232.
  arrangement, 247, 252.
  framing, 233.
Round dairy barn, 247, 252.
Rubble masonry, 19.

Sand, 32.
Sapwood, 2.
Seasoning, 7.
Septic tanks, 338.
Sewage disposal, 336.
Sheathing, 92.
Sheep barns, 212.
  cost, 213.
  design, 213.
  essentials, 212.
  ventilation, 214.
Sheet glass, 62.
Shingles, 29.

slate, 29.
wood, 29.
Shop, 230.
Shrinkage of timber, 88.
Silage,
  pressure, 140.
  weight, 141.
Silos, 136.
  capacities, 139.
  chute, 182.
  definition, 136.
  development, 137.
  doors, 146, 156, 176.
  essentials, 137.
  foundation, 143, 159.
  size, 138.
  types, 141.
    concrete, 156.
    gurler, 149.
    pit, 185.
    stave, 142.
    vitrified tile, 151.
Sinks, 335.
Siphon-jet closet, 334.
Slate shingles, 29.
Smith swine house, 211.
Solvent, 57.
Springwood, 3.
Squared-stone masonry, 19.
Stack, 301.
Stairs, 102.
  Boston, 104.
  construction, 104.
  definitions, 103.
  dimensions, 104.
  housed, 105.
Stallion barn, 255.
Star-shake, 6.
Stave silo, 142.
Steam heating, 302.
Steel, 15.
  classification, 15.
  manufacture, 15.
  properties, 16.
Stone,
  building, 21.
  classification, 16.
  cutting, 20.
  varieties of building, 22.
Stone masonry, 18.
  varieties, 18.
Storage barns, 218.
Stoves, 300.

Strength,
  concrete, 51.
  joists and girders, 226.
  steel, 16.
  wood, 12.
Stucco, 41.
  application, 42.
  constituents, 42.
  finishing, 45.
Summerwood, 3.
Swine houses, 202.
  cost, 204.
  equipment, 204.
  types, 205.

Tie, brick, 27.
Tile, roofing, 30.
Timber, 1.
Timber framing, 219.
Tin roofing, 30.
Traps, 333.
Truss over opening, 87.
Truss, scissors, 125.
Twisted, 7.

Upset, 7.

Vacuum heating system, 311.
Varnish, 59.
  application, 60.
  definition, 59.
Vehicle, 56.
Veneer, 10.
Ventilating fireplace, 278.
Ventilation, 273.
  definition, 274.
  farm building, 280.
  fireplace, 278.
  King system, 280.
  purpose, 274.
  residence, 274.
  stove-heated buildings, 278.
Vitrified tile, 152.
Vitrified-tile silo, 151.

Walls, 92.
  foundation, 82.
  residence, 92.
  silo, 138.
Washdown closet, 334.
Washout closet, 334.
Water back, 328.
Water closets, 338.
  hopper, 334.
  siphon-jet, 334.
  washdown, 334.
  washout, 334.
Water hammer, 307.
Water heater, gas, 328.
Waterproofing concrete, 37.
Water-supply systems, 317.
  design of hydro-pneumatic, 322.
  gravity tank, 321.
  hydraulic ram, 318.
  hydro-pneumatic, 321.
  pneumatic, 324.
  types, 317.
Wells, 315.
  artesian, 317.
Wet rot, 6.
Windows, 94.
Wind-shake, 6.
Wire nails, 63, 66.
Wire straightener, 155.
Wisconsin model dairy barn, 244.
Wood, 1.
  color, 4.
  crushing strength, 12.
  defects, 5.
  kiln-drying, 8.
  odor, 5.
  seasoning, 7.
  shearing strength, 12.
  structure, 1.
  tensile strength, 12.
  testing, 11.
  varieties, 13.
Worms, 5.
Wrought nails, 63.